Current Developments in Mathematics 2011

edited by

David Jerison
Barry Mazur
Tomasz Mrowka
Wilfried Schmid
Richard Stanley
Shing-Tung Yau

International Press
www.intlpress.com

Current Developments in Mathematics, 2011

2010 Mathematics Subject Classification. 05A15, 05C50, 26C10, 30C15, 34B45, 60B20, 60F05, 60H15, 60K35, 62H20, 82C22, 82C44.

ISBN 978-1-57146-239-8

Typeset using the LaTeX system.

Contents

Current Developments in Mathematics, 2011

The Laplacian on planar graphs and graphs on surfaces

R. Kenyon*

ABSTRACT. We discuss the Laplacian operator on graphs embedded on a surface, concentrating on simple surfaces (disk, annulus, torus). In particular we review and extend results on the reconstruction of the network from boundary measurements, connections with the spanning tree and grove model, and the characteristic polynomial of the Laplacian on an annulus and torus.

CONTENTS

1991 *Mathematics Subject Classification.* 05C50, 34B45.

*Brown University, 151 Thayer St., Providence, RI 02912. Research supported by the NSF.

1. Introduction

By a **resistor network**, or simply **network**, we mean here a graph $\mathcal{G} = (V, E)$ with vertices V and (undirected) edges E, and c a positive real-valued function on the edges. The value c_e is the **conductance** of the edge e.

The **Laplacian** on a network $(\mathcal{G}, c)$ is the linear operator $\Delta\colon \mathbb{R}^V \to \mathbb{R}^V$ defined by

$$(\Delta f)(v) = \sum_{v' \sim v} c_{vv'}(f(v) - f(v')). \tag{1}$$

Here the sum is over neighbors v' of v.

The Laplacian is the most basic operator on a network, and has countless uses in all areas of mathematics. We study here some specific properties of the Laplacian on networks embedded on surfaces, and in particular on the simplest surfaces: the plane, the annulus, the torus.

Our goals will be to focus on a few topological, probabilistic and combinatorial applications. Specifically, our three main interconnected goals are to discuss:

(1) The discrete EIT (Electrical Impedance Tomography) problem: reconstructing the network from boundary measurements. For planar networks, the classification due to Yves Colin de Verdière of Dirichlet-to-Neumann operators [**6, 7**],
(2) The connections with the random spanning tree model developed by Curtis, Ingerman and Morrow [**8**] and Wilson and myself [**19, 20**], and
(3) The characteristic polynomial of Δ on an annulus and torus based on joint work with Okounkov and Sheffield [**16**], Okounkov [**15**], and in [**14**].

Acknowledgements. Large parts of Section 4 were developed by Colin de Verdière [**6**] and Curtis, Ingerman and Morrow [**8**]. I thank them for explanations and discussions.

The rest of this paper includes material developed in various projects jointly with Alexander Goncharov, Andrei Okounkov, Scott Sheffield, and David Wilson. In particular some of the material is lifted directly out of work in progress with David Wilson [**21**], whom I thank for allowing it to be included here.

The main new results in the current paper are Lemma 4.11 and Theorems 5.1, 6.1, and 7.1.

2. Background

2.1. Electrical Impedance Tomography. The classical **Calderón problem** is to determine the electrical conductivity of a region in $\mathbb{R}^2$ or $\mathbb{R}^3$ by making boundary measurements of the type: fix the voltage along the boundary and measure the resulting current flow out of the region.

This problem has concrete applications in medical imaging, nondestructive materials testing, etc. In these settings it is called Electrical Impedance Tomography (EIT). It is widely used because it is easy and cheap to implement:

basically it is very easy to apply a low voltage to pretty much any conductive material and measure the resulting current flow.

In this paper we study the *discrete version* of EIT. Given a resistor network (a finite graph in which every edge is a resistor), understand its internal structure from boundary measurements. There is a very nice solution for planar networks. For networks on more general surfaces we are beginning to have a better understanding but much more remains to be done.

For a network $\mathcal{G}$ on a surface with nontrivial topology there is closely related question, even when there is no boundary, which is to understand the determinant of the "bundle Laplacian" as a function on the moduli space of flat connections on bundles on $\mathcal{G}$.

We shall see how these questions are related below.

2.2. Spanning trees. One of the combinatorial tools involved in the study of discrete EIT is the model of (random) **spanning trees**, and related objects called **cycle-rooted spanning forests** on networks. Kirchhoff [**22**] was the first to see the connection between spanning trees and the Laplacian on a network: he showed that the Laplacian determinant in fact counts spanning trees.

The random spanning tree measure (or UST, for Uniform Spanning Tree) has for the past 20 years been a remarkably successful and rich area of study in probability theory. The random spanning tree as a probability model goes back to at least work of Temperley [**31**], who gave a bijection between spanning trees on the square grid and domino tilings. Aldous and Broder [**1, 2**] in the 1980s showed a connection between spanning trees and random walk, giving an algorithm for sampling a UST based on the simple random walk on the network. Pemantle [**28**] showed that the unique path between two points in a UST has the same distribution as the **loop-erased random walk** (LERW) between those two points[1]. Wilson [**32**] extended this to give a quick and elegant method of sampling a uniform spanning tree in any graph. Burton and Pemantle [**4**] proved that the edges of the uniform spanning tree form a "determinantal process", that is there is a matrix T indexed by the edges such that the probability that some set of edges $\{e_1, \dots, e_k\}$ is in the UST is $\det(T(e_i, e_j)_{1\le i,j\le k})$. The matrix T is built from the Green's function of Δ.

The UST on the grid $\mathbb{Z}^2$ received particular attention due to the conformal invariance properties of its scaling limit (limit when the mesh size of the grid goes to zero). In particular in [**13**] we showed that the "winding field" of the UST on $\mathbb{Z}^2$ (which measure the winding of the branches around faces) converges in the scaling limit to the Gaussian free field, a conformally invariant random function on $\mathbb{R}^2$. In [**25**] Lawler, Schramm and Werner showed the convergence of the branches of the UST on $\mathbb{Z}^2$ in the scaling limit to the conformally invariant process SLE_2.

[1]The loop-erased random walk between a and b is defined as follows: talk a simple random walk from a stopped when it reaches b, and then erase from the trace all loops in chronological order. What remains is a simple path from a to b.

As mentioned above, Temperley [**31**] gave a bijection between spanning trees and the dimer model on the square grid; this was later generalized to arbitrary graphs in [**17**] and [**18**], pointing to a very close relation between the UST and the dimer model. Many of the results discussed below are inspired by the corresponding results in the dimer model.

2.3. Characteristic polynomials. To a network on an annulus we can associate a polynomial $P(z)$ which is the determinant of the line-bundle Laplacian Δ for a flat line bundle with monodromy $z \in \mathbb{C}^*$ around a generator of the homology of the annulus. One can think of this as the determinant of the Laplacian acting on the space of z-quasiperiodic functions: "locally defined" functions whose values are multiplied by z under analytic continuation counterclockwise around the annulus.

The polynomial P is real and symmetric (that is, $P(z) = P(1/z)$). There is a nice characterization of those P which occur: they are those with positive constant coefficient, and whose roots are real, positive and distinct except for a double root at $z = 1$. See Theorem 6.1 below.

For a network on the torus one can associate a two-variable polynomial $P(z_1, z_2)$ in a similar way: it is the determinant of the Laplacian for a flat line bundle with monodromy z_1, z_2 around generators of π_1 of the torus. This polynomial plays an important role in understanding the structure and large-scale properties of the random spanning tree model on a periodic planar graph [**4, 16**]. Like in the case of the annulus there is a characterization of those $P(z_1, z_2)$ which occur: the curve $\{P(z_1, z_2) = 0\}$ is a *Harnack curve* which is symmetric ($P(z_1, z_2) = P(z_1^{-1}, z_2^{-1})$), satisfies $P(1, 1) = 0$ and has positive constant coefficient. See the definition of Harnack curve in Section 7.4. One can show that in fact any (simple) Harnack curve satisfying the above properties arises from the characteristic polynomial of some toral network.

A more recent development in networks on the torus is the understanding of the integrable dynamical system underlying them. Here the dynamical system is not related to the flow of current in the network, but is a dynamical system on the space of conductances assigned to the edges in the network. It is the restriction of an integrable Hamiltonian dynamical system on a larger space (the space of line bundles on bipartite networks on the torus) to a Lagrangian subvariety. See [**12**]. We will not discuss this here, mostly because we don't yet have a very good understanding of this system except as an invariant subvariety of a larger system.

For networks on other surfaces Σ one can also define a characteristic polynomial $P = \det \Delta$ which is a regular function on the representation variety of $\pi_1(\Sigma)$ into $SL_2(\mathbb{C})$. Here much less is known about the structure of P. See however Theorem 8.1 below.

Finally, one would like to consider a network on a general surface with boundary, and understand properties of the boundary measurement map (for the appropriate bundle Laplacian) and to what extent it determines the network. We are still far from this goal but the cases of simple topology or small boundary hint that there is an interesting answer.

3. Networks

By a **network** we mean a connected finite graph $\mathcal{G} = (V, E)$ with a positive real-valued function c on the edges called the **conductance** function, and a usually nonempty set of vertices $\mathcal{N}$ called **nodes** or **terminals** or **boundary vertices**. These are the simplest type of electrical networks, with resistors only.

The nodes are indexed from 1 to n. Let I be the set of non-node vertices; elements of I are referred to as **internal vertices**.

3.1. Laplacian. The Laplacian on a network $\mathcal{G}$ is the operator $\Delta : \mathbb{R}^V \to \mathbb{R}^V$ defined for $f \in \mathbb{R}^V$ by (1). Since $\mathbb{R}^V$ has a natural basis, consisting of functions δ_v which are 1 at v and zero elsewhere, we often write Δ as a matrix $\Delta = (\Delta_{vv'})_{v,v' \in V}$ where $\Delta_{vv'} = -c_{vv'}$ if v, v' are adjacent and the diagonal entries are $\Delta_{vv} = \sum_{v' \sim v} c_{vv'}$. See for example (8) which is the Laplacian of the graph of Figure 1.

Associated to the Laplacian is the quadratic form Q on $\mathbb{R}^V$ defined as

$$Q(f) = \langle f, \Delta f \rangle = \sum_{v \sim v'} c_{vv'}(f(v) - f(v'))^2 \tag{2}$$

which is called the **Dirchlet energy** of f. Here $\langle , \rangle$ is the natural inner product in which the basis $\{\delta_v\}_{v \in V}$ is orthonormal.

From the expression (2) we see that Δ is positive semidefinite with kernel consisting of the constant functions (we assume unless stated otherwise that $\mathcal{G}$ is connected).

Another useful operator is the **incidence matrix** d, defined as follows. Fix an (arbitrary) orientation for each edge e, so that e_+ and e_- are its

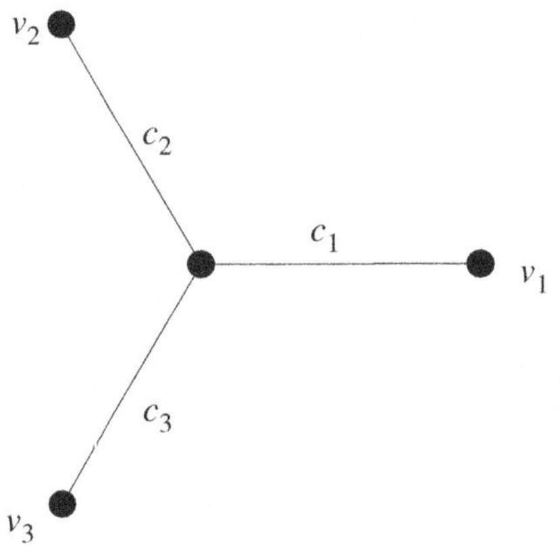

FIGURE 1. The 'Y' network.

head and tail. Define $d\colon \mathbb{R}^V \to \mathbb{R}^E$ by the formula

$$df(e) = f(e_+) - f(e_-).$$

Then it is easy to verify that

$$\Delta = d^* \mathcal{C} d,$$

where $\mathcal{C}$ is the diagonal matrix of conductances, and d^* is the transpose of d (its adjoint for the inner product $\langle , \rangle$ above): for an element $w \in \mathbb{R}^E$ we have

$$d^* w(v) = \sum_{e \to v} w(e)$$

where the sum is over the edges e containing endpoint v, and directed towards v. We can call df the **gradient** of the function f and d^*w the **divergence** of the flow w.

Note that we can write the Dirichlet energy as

$$Q(f) = \langle df, \mathcal{C} df \rangle$$

where here $\langle , \rangle$ is the natural inner product on $\mathbb{R}^E$.

3.2. The Laplacian with boundary. If $B \subset V$ is a subset of vertices and $I = V \setminus B$, we can define an operator $\Delta_B : \mathbb{R}^I \to \mathbb{R}^I$ called the **Dirichlet Laplacian with boundary** B by the same formula

$$\Delta_B f(v) = \sum_{v' \sim v} c_{vv'}(f(v) - f(v')),$$

where here $v \in I$ and the sum is over all neighbors of v in V (not just those in I).

As a matrix Δ_B is just a submatrix of the full Laplacian Δ: the submatrix indexed by I.

3.3. The Dirichlet problem. A function $f \in \mathbb{R}^V$ is said to be **harmonic at** v if $\Delta f(v) = 0$. By (1) this simply means that $f(v)$ is a weighted average of the values of f at the neighboring vertices, where the weights are proportional to the conductances.

A function is **harmonic** if it is harmonic at all vertices. On a finite connected graph a harmonic function must be constant, since the kernel of Δ consists of constant functions only.

Suppose B is a nonempty set boundary vertices and u is a function on B. A function f on V which is harmonic on $I = V \setminus B$ and agrees with u on B is said to be **harmonic with Dirichlet boundary conditions on** B. Such functions satisfy a **maximum principle**: their maximum (and minimum) values occur on B (because of the above observation about weighted averages).

Given this fact it is not hard to see that, if B is nonempty, then Δ_B is invertible: any function f in the kernel of Δ_B can be extended to a function on V taking values zero on B; it is thus a harmonic function with Dirichlet boundary conditions at B. By the maximum principle it must take its maximum and minimum on B, where it has value 0. Thus f must be identically zero.

The **Dirichlet problem** is the problem of finding, for a given function u on B, a function f on V which is equal to u on B and harmonic on $I = V \setminus B$: find f satisfying

$$\begin{aligned} \Delta f(v) &= 0 \quad \text{for } v \in V \setminus B \\ f(v) &= u(v) \quad \text{for } v \in B. \end{aligned} \tag{3}$$

The function f is called the **harmonic extension** of u. For finite networks the solution to the Dirichlet problem is straightforward using invertibility of Δ_B. Alternatively, the solution f is the function with boundary values u and which minimizes the Dirichlet energy $Q(f)$. Convexity of Q implies that there is a unique solution.

A function on $\mathcal{G}$ which is harmonic on I is a solution to the Dirichlet problem with boundary B. Such a function is also called an **equilibrium potential**.

3.4. The Dirichlet problem and the response matrix. For a network we will take $\mathcal{N}$, the set of nodes, to play the role of the boundary vertices (which we called B above). Given a function u on $\mathcal{N}$, let f_u be the harmonic extension of u. Then $\Delta f_u(v)$, restricted to $\mathcal{N}$, is a (generally nonzero) function of $v \in \mathcal{N}$. The map $\Lambda : u \mapsto \Delta f_u|_{\mathcal{N}}$ is a linear map $\Lambda : \mathbb{R}^{\mathcal{N}} \to \mathbb{R}^{\mathcal{N}}$ called the **Dirichlet-to-Neumann map** of the network. For certain conventional reasons we often prefer to work with $L = -\Lambda$ and refer to L as the **response matrix**. See the next section for an example.

We can describe Λ as a matrix as follows. Suppose we index the vertices so that vertices in $\mathcal{N}$ come first. Then we can write Δ in block form as

$$\Delta = \begin{pmatrix} A & B \\ B^t & C \end{pmatrix},$$

where A has rows and columns indexed by the nodes $\mathcal{N}$ and C has rows and columns indexed by the internal vertices; C is just the matrix of the Dirichlet Laplacian $\Delta_{\mathcal{N}}$.

The Dirichlet problem (3) can be written as follows: find f_I on the internal vertices I so that

$$\begin{pmatrix} A & B \\ B^t & C \end{pmatrix} \begin{pmatrix} u \\ f_I \end{pmatrix} = \begin{pmatrix} g \\ 0 \end{pmatrix},$$

for a function g which is the desired function of u. We can split this into two linear equations

$$Au + Bf_I = g \tag{4}$$

$$B^t u + Cf_I = 0. \tag{5}$$

This second equation can be solved as

$$f_I = -C^{-1}B^t u. \tag{6}$$

(Here C is invertible, as discussed above).

One can then plug in this value of f_I in the first equation to get

$$Au - BC^{-1}B^t u = g,$$

that is, $\Lambda u = g$ where Λ is the matrix $\Lambda = A - BC^{-1}B^t$. Thus

$$L = -A + BC^{-1}B^t. \tag{7}$$

We see from this expression that L is symmetric, since A and C are symmetric. In fact Λ is positive semidefinite (and L is negative semidefinite), with kernel consisting solely of the constant functions. This follows from the fact that

$$\langle u, \Lambda u\rangle = \langle f, \Delta f\rangle \geq 0$$

where f is the harmonic extension of u (one can check this using (6)).

3.4.1. *Example.* Let $\mathcal{G}$ be the network of Figure 1 with three nodes $\{v_1, v_2, v_3\}$, one internal vertex v_4 and three edges with conductances c_1, c_2, c_3. We have

$$\Delta = \begin{bmatrix} c_1 & 0 & 0 & -c_1 \\ 0 & c_2 & 0 & -c_2 \\ 0 & 0 & c_3 & -c_3 \\ -c_1 & -c_2 & -c_3 & c_1+c_2+c_3 \end{bmatrix}. \tag{8}$$

Using formula (7) we have

$$L = -\begin{pmatrix} c_1 & 0 & 0 \\ 0 & c_2 & 0 \\ 0 & 0 & c_3 \end{pmatrix} + \begin{pmatrix} -c_1 \\ -c_2 \\ -c_3 \end{pmatrix} \left(\frac{1}{c_1+c_2+c_3}\right) \begin{pmatrix} -c_1 & -c_2 & -c_3 \end{pmatrix}$$

$$= \begin{pmatrix} \frac{-c_1c_2 - c_1c_3}{c_1+c_2+c_3} & \frac{c_1c_2}{c_1+c_2+c_3} & \frac{c_1c_3}{c_1+c_2+c_3} \\ \frac{c_2c_1}{c_1+c_2+c_3} & \frac{-c_2c_1 - c_2c_3}{c_1+c_2+c_3} & \frac{c_2c_3}{c_1+c_2+c_3} \\ \frac{c_3c_1}{c_1+c_2+c_3} & \frac{c_3c_2}{c_1+c_2+c_3} & \frac{-c_3c_1 - c_3c_2}{c_1+c_2+c_3} \end{pmatrix}. \tag{9}$$

3.5. Electrical interpretation. Associated to a function f (called **potential** or **voltage**) on a network is a flow ω on the edges (called **current**) defined by

$$\omega(e) = Cdf(e) = c_e(f(e_+) - f(e_-)).$$

Here C is the diagonal matrix of edge conductances and $df(e) = f(e_+) - f(e_-)$ is the gradient of f.

Ohm's law states that $\mathcal{C}df(e)$ is the flow across edge e when the vertices are held at potentials $f(e_+)$ and $f(e_-)$, and the **resistance** on the edge is $1/c_e$, the reciprocal of the conductance.

Kirchhoff's network equations state that when the boundary nodes are held at fixed potentials, the internal vertices will attain potentials f in such a way that the flow of current induced by Ohm's law in the network has the property that the current entering an internal vertex equals the current exiting that vertex. In other words the divergence $d^*\omega$ of the current ω is zero at internal vertices. Thus $d^*\mathcal{C}d(f) = 0$ at every internal vertex, so f is the harmonic extension of u.

This allows us to give L_{ij} the following electrical interpretation: Hold all nodes except i at potential 0, and node i at potential 1. Then for the resulting current flow, L_{ij} is the current exiting the network at node i. This current is positive: the potentials at internal vertices are in $[0, 1]$ (in fact in $(0, 1)$) by the maximum principle for harmonic functions, so the current at edges adjacent to j is directed towards j. Similarly one shows $L_{ii} < 0$.

4. Circular planar networks

By a **circular planar network** (CPN) we mean a connected finite graph $\mathcal{G} = (V, E)$ embedded in the plane, with a positive real-valued function c on the edges called the **conductance** function, and a nonempty set of vertices $\mathcal{N}$ on the outer face called **nodes** or **terminals**. (There may be other vertices on the outer face as well). The reason it is a *circular* planar network is that the network can be drawn inside the disk with the nodes on the boundary circle.

The nodes are ordered in counterclockwise order from 1 to n. Let I be the set of non-node vertices; elements of I are referred to as **internal vertices**. The non-exterior faces of $\mathcal{G}$ are referred to as **interior faces**; we consider $\mathcal{G}$ to also have n **exterior faces**, with the ith exterior face comprising the region between nodes i and $i + 1$, and having boundary consisting of the edges along the "outer face" of $\mathcal{G}$ between node i and $i + 1$.

4.1. Conjugate harmonic function. If $\mathcal{G}$ is a circular planar network, a harmonic function f on $\mathcal{G}$ with Dirichlet boundary conditions on $\mathcal{N}$ has a **conjugate harmonic function** g defined (up to an additive constant) on the faces of $\mathcal{G}$ by the condition $\mathcal{C}df = \partial g$, where ∂ is the difference operator on the dual graph $\mathcal{G}'$: if $e = vv'$ is an edge of $\mathcal{G}$ and F_1, F_2 are the adjacent faces, so that F_1 is left of e when e is traversed from v to v', then

$$c_e(f(v') - f(v)) = g(F_1) - g(F_2). \tag{10}$$

The harmonicity of f implies that these equations are consistent around a vertex; by summing along paths one sees that the value of g is defined consistently on all faces, once one choose a value of g at any starting face. Thus these equations define g uniquely up to a global additive constant. The function g is harmonic for the operator $\partial^*\mathcal{C}^{-1}\partial$. That is, one puts conductances on the dual edges which are the inverses of the conductances on the primal edges.

The pair (f, g) is sometimes written as $f + ig$ and is called a **discrete analytic function** since the equations (10) play the role of the discrete Cauchy-Riemann equation.

If f is harmonic with Dirichlet boundary conditions then g will be harmonic on $\mathcal{G}'$ with Dirichlet boundary conditions, that is, g is harmonic on interior vertices of $\mathcal{G}'$ (that is, on non-exterior faces of $\mathcal{G}$).

4.2. Characterization of response matrices. The response matrix is a fundamental matrix associated to a network. A natural question is: what properties of the network are reflected in its response matrix?

The basic properties of L shown in the previous section are that

(1) L is real and symmetric,
(2) L is negative semidefinite,
(3) The kernel of L consists of the constant functions, and
(4) $L_{ij} > 0$ if $i \neq j$, and $L_{ii} < 0$.

(It is not hard to see that property 2 is implied by the others).

Any matrix satisfying the above properties is in fact the response matrix of some (not necessarily planar) network: it suffices to take a complete graph on n vertices, declare that each vertex is a node, and put conductance L_{ij} on the edge ij connecting vertices i and j.

The question is much more interesting in the case of circular planar networks. In this case there is a beautiful characterization of response matrices in terms of minors, due to Colin de Verdière [**6**].

Given two disjoint subsets A, B of the nodes of a circular planar network, we say that A and B are **noninterlaced** if they are contained in disjoint intervals of nodes with respect to the circular order. That is, if $a < b < c < d$ are nodes then it is never the case that $a, c \in A$ and $b, d \in B$.

THEOREM 4.1 ([**6**]). *L is the response matrix of a connected circular planar network if and only if L is symmetric, with kernel consisting of the constant functions, and for any disjoint noninterlaced subsets of nodes A and B with $|A| = |B|$, we have* $\det L_A^B \geq 0$, *where L_A^B is the submatrix of L whose rows index A and columns index B.*

There are two parts of this theorem, the necessity of the conditions and the existence of a network for any L satisfying the conditions. Both have interesting proofs which we will discuss.

We define a network to be **well-connected** if all noninterlaced minors $\det L_A^B$ are *strictly* positive. In [**6**] Colin de Verdière shows that the space of well-connected circular planar networks on n nodes is homeomorphic to an open ball of dimension $\binom{n}{2}$. More about this in Section 4.4.2.

4.3. Groves. As one can see from (7), the response matrix entries are rational functions of the conductances. Surprisingly, the off-diagonal entries are **subtraction free** rational functions of the conductances, that is, ratios of polynomials with positive coefficients; in fact the coefficients are all 1 or 0 (see for example (9)). This is a strong indication that these polynomials are counting some type of combinatorial objects.

Indeed, that is one of the ideas in the Curtis-Ingerman-Morrow proof (see [**8**]) of necessity in Theorem 4.1 above, which we discuss in this section.

4.3.1. *Spanning trees.* A **spanning tree** of a graph is a subset of edges which is a tree, that is, contains no cycles, and is spanning, that is, connects all vertices. Equivalently one can say that a spanning tree is a maximal set of edges (with respect to inclusion) that has no cycles.

For a network $\mathcal{G}$ with conductances c on its edges, one can define the **weight** $w(T)$ of a spanning tree T to be the product of the conductances of its edges, $w(T) = \prod_{e\in T} c_e$.

The most important theorem about spanning trees is that the Laplacian determinant is the weighted sum of spanning trees (this is also called the **partition sum**).

THEOREM 4.2 (Matrix-Tree Theorem [**22**]). *Let* $\det_0 \Delta$ *be the product of the nonzero eigenvalues (with multiplicity) of* Δ *for a connected network* $\mathcal{G}$. *Then*

$$\frac{1}{|\mathcal{G}|}\det_0\Delta = \sum_T w(T).$$

Here the sum is over spanning trees of $\mathcal{G}$.

This theorem is usually attributed to Kirchhoff [**22**]. An equivalent formulation is that the partition sum $Z := \sum_T w(T)$ is equal to the **reduced determinant** of Δ, which is the determinant of Δ after removing any row and column; since Δ is symmetric and constant functions are in the kernel any choice of row and column gives the same reduced determinant.

PROOF OF THEOREM 4.2, SKETCH. Several proofs of the Matrix-tree Theorem are known; the classical one we present uses the **Cauchy-Binet Theorem** which is a formula for computing the determinant of a product of two nonsquare matrices: it states that if X is an $n\times m$ matrix and Y is $m\times n$ with $m\ge n$ then

$$\det XY = \sum_B \det(X^B)\det(Y_B) \tag{11}$$

where B runs over all choices of n column indices for X, X^B is the $n\times n$ minor of X defined by columns B and Y_B is the $n\times n$ minor of Y defined by rows B.

For the Matrix-Tree Theorem, let $\Delta_{\bar{1},\bar{1}}$ be the submatrix of Δ obtained by removing row 1 and column 1. Then $\Delta_{\bar{1},\bar{1}} = d_{\bar{1}}^* C d_{\bar{1}}$ where $d_{\bar{1}}$ is obtained from d by removing column 1.

One now uses $X = d_{\bar{1}}^*$ and $Y = Cd_{\bar{1}}$ in (11). Each B in (11) is a set of $|V|-1$ edges; we claim that the nonzero summands in (11) are precisely the sets of edges which form spanning trees. If B forms a tree, its edges can be directed towards the removed vertex. Then in the definition of $\det d_{\bar{1},B}$ as an expansion over the symmetric group,

$$\det d_{\bar{1},B} = \sum_{\sigma \in S_{n-1}} (-1)^{\sigma} d_{2\sigma(2)} \dots d_{n\sigma(n)},$$

the only nonzero term is the one whose permutation σ matches each vertex to the edge adjacent to it and closer to the root. Thus $\det d_{\bar{1},B} = \pm 1$. If edges in B do not form a tree, then there must be more than one component; the function on vertices which is 1 on a component not containing the removed vertex and zero elsewhere is in the kernel of $d_{\bar{1},B}$, so $\det d_{\bar{1},B} = 0$.

Finally, the determinant of the diagonal submatrix $\det \mathcal{C}_B^B$ for a tree B is the product of the edge weights, and $\det d^*_{\bar{1},B} = \det d_{\bar{1},B} = \pm 1$. So

$$\det(d^*_{\bar{1},B} \mathcal{C}_B^B d_{\bar{1},B}) = w(T).$$

This completes the proof. □

The **spanning tree measure** is the probability measure on spanning trees of a graph in which the probability of a tree is proportional to its weight. In the special case that the conductances are all 1 this is the **uniform measure on spanning trees** (UST measure); each tree has probability $\frac{1}{\kappa}$ where κ is the determinant of the reduced Laplacian of the conductance-1 network; κ is sometimes called the **complexity** of the network.

Cayley's formula [**5**], first proved by Borchhardt [**3**] gives the number spanning trees of the complete graph K_n to be n^{n-2}. This follows easily from the above theorem since

$$\Delta = \begin{pmatrix} n-1 & -1 & \dots & -1 \\ -1 & n-1 & -1 & \vdots \\ \vdots & & \ddots & -1 \\ -1 & \dots & -1 & n-1 \end{pmatrix}$$

whose eigenvalues are n with multiplicity $n-1$ and 0 with multiplicity 1.

4.3.2. *Groves from minors of L.* On a network with nodes $\mathcal{N}$, a **grove** is a subset of the edges, each of whose components is a tree containing one or more nodes. In other words it is a "spanning forest" in which each component is connected to the boundary (and isolated vertices are considered components, so there are no isolated internal vertices). See Figure 2. We allow a component to contain more than one node; for example a spanning tree is a special case of a grove.

Groves can be put into equivalence classes depending on how their components partition the nodes. Given a partition π of the nodes, let S_π be the set of groves in which each component contains exactly all the nodes of a part of π. For example when there are three nodes the partitions are: $1|2|3$ where each node is in a separate component, $1|23$ where nodes 2 and 3 are

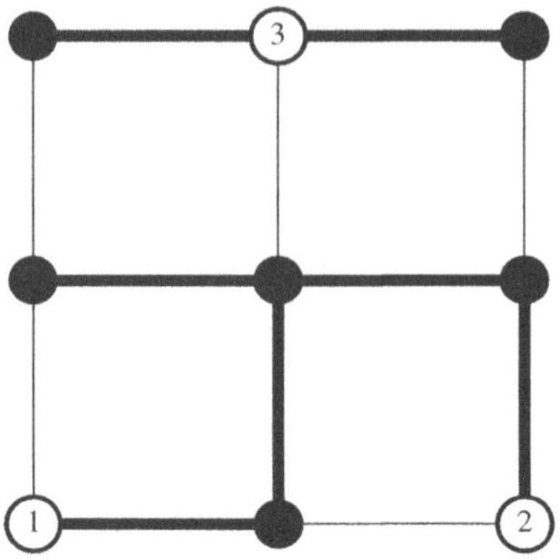

FIGURE 2. A grove of type 12|3 (nodes are in white; internal vertices in black).

in the same component different from 1, and likewise $2|13, 3|12$, and finally 123 where all nodes are in the same component. More generally groves in the set $S_{1|2|\dots|n}$, that is, in which each component contains exactly one node, are called **uncrossings**. The groves in $S_{12\dots n}$ are spanning trees.

Note that for circular planar graphs not all partitions π are possible; for a partition to be obtained by a grove it must be planar. For example 13|24 is not the partition of any grove on a circular planar graph.

The probability measure on spanning trees extends to a probability measure on groves: the weight of a grove is the product of its edge weights, and the probability of a grove is $\frac{1}{Z}$ times its weight, where $Z = \sum_T w(T)$ is the weighted sum of all groves.

For each partition π we denote by Z_π the weighted sum of groves of type π. Then $Z = \sum_\pi Z_\pi$, and the probability that a random grove has type π is $\frac{Z_\pi}{Z}$. We define $Z_{\text{unc}} = Z_{1|2|\dots|n}$ to be the partition sum for uncrossings, and $Z_{\text{tree}} = Z_{12\dots n}$ to be the weight sum of spanning trees.

All minors of the response matrix have combinatorial interpretations in terms of groves, for example when $i \neq j$ we prove in Theorem 4.3 below that

$$L_{ij} = \frac{Z_{\pi(ij|\text{rest different})}}{Z_{\text{unc}}} = \frac{\Pr(\pi(ij|\text{rest different}))}{\Pr(\text{unc})}$$

where $\pi(ij|\text{rest different})$ is the partition in which all nodes are in separate components except for i and j which are in the same component. More generally, we have the following combinatorial interpretation of *all* minors of L (a closely related result can be found in [**8**]):

THEOREM 4.3 ([**20**]). *Let $Q, R, S, T \subseteq \mathcal{N}$ be a partition of $\mathcal{N}$ into four sets with $|R| = |S|$ (and some of which may be empty). Then $\det L_{RUT}^{SUT}$ is the ratio of two terms: the denominator is Z_{unc}; the numerator is a signed weighted sum of groves of $\mathcal{G}_T$, the graph $\mathcal{G}$ in which all nodes in T are considered internal vertices, the nodes in Q are in singleton parts, and in* *which nodes in R are paired with nodes in S, with the sign being the sign of the pairing permutation:*

$$\det L_{R\cup T}^{S\cup T} = (-1)^{|T|} \sum_{\text{permutations } \rho} (-1)^{\rho} \frac{Z\left[\begin{smallmatrix} s_{\rho(1)} \\ r_1 \end{smallmatrix}\Big|\cdots\Big|\begin{smallmatrix} s_{\rho(k)} \\ r_k \end{smallmatrix}\Big|q_1\Big|\cdots\Big|q_\ell\right]}{Z_{\mathrm{unc}}}. \tag{12}$$

As a special case of this theorem, we have the above statement about L_{ij} for $i \neq j$ (here L_{ij} is the i,j-entry of L which in the notation of the theorem would be $L_{\{i\}}^{\{j\}}$). Another special case is the reduced determinant of L: for any i and j, removing row i and column j gives

$$\det L_{\bar{i}}^{\bar{j}} = \frac{Z_{\mathrm{tree}}}{Z_{\mathrm{unc}}}$$

(to see this, take $R = \{i\}, S = \{j\}$ and $T = \mathcal{N}\backslash\{i,j\}$).

In the example (9) above where $\mathcal{G}$ is the Y network of Figure 1, $L_{12} = \frac{c_1c_2}{c_1+c_2+c_3}$; c_1c_2 is the weight of the unique grove of type $12|3$ and $c_1+c_2+c_3$ is the sum of the weights of the three groves of type $1|2|3$.

Proof of Theorem 4.1, necessity. Since nonplanar pairings do not occur, if R and S are non-interlaced subsets of the nodes of the same size, then $\det L_R^S$ is a nonnegative because there is a single term in the sum in (12) and the permutation is the identity. This gives a combinatorial interpretation of the inequalities in Theorem 4.1: they are positive because they count something. □

Proof of Theorem 4.3. Write Δ in block form as

$$\Delta = \begin{pmatrix} A & B \\ B^t & C \end{pmatrix}$$

where A is indexed by the nodes and C by the internal vertices. Then we can factor Δ as

$$\Delta = \begin{pmatrix} A & B \\ B^t & C \end{pmatrix} = \begin{pmatrix} A - BC^{-1}B^t & BC^{-1} \\ 0 & I \end{pmatrix}\begin{pmatrix} I & 0 \\ B^t & C \end{pmatrix}$$
$$= \begin{pmatrix} -L & BC^{-1} \\ 0 & I \end{pmatrix}\begin{pmatrix} I & 0 \\ B^t & C \end{pmatrix}.$$

We wish to compute the determinant of the minor $\Delta_{X\cup I}^{Y\cup I}$. For notational simplicity reorder the nodes so that nodes in X come last and nodes in Y come first. Then

$$\Delta = \begin{pmatrix} * & * & * \\ -L_X^Y & * & * \\ 0 & 0 & I \end{pmatrix}\begin{pmatrix} I & 0 & 0 \\ 0 & I & 0 \\ * & * & C \end{pmatrix}.$$

and

$$\Delta_{X\cup I}^{Y\cup I} = \begin{pmatrix} -L_X^Y & * & * \\ 0 & 0 & I \end{pmatrix}\begin{pmatrix} I & 0 \\ 0 & 0 \\ * & C \end{pmatrix}.$$

The Cauchy-Binet formula (11) then gives

$$\det \Delta_{X\cup I}^{Y\cup I} = (-1)^{|X|} \det L_X^Y \det C.$$

In particular

$$\det \Delta_{R\cup T\cup I}^{S\cup T\cup I} = (-1)^{|R|+|T|} \det L_{R\cup T}^{S\cup T} \det C$$

where I is the set of internal nodes. Since $Z_{\text{unc}} = \det C$, it suffices to show that $\det \Delta_{R\cup T\cup I}^{S\cup T\cup I}$ is the desired numerator.

Let $I' = I \cup T$ be the "new" set of internal nodes. We need to evaluate $\det \Delta_{R\cup I'}^{S\cup I'}$. The remainder of the proof is now an extension of the proof of Theorem 4.2. Write $\Delta = d^*\mathcal{C}d$. Then $\Delta_{R\cup I'}^{S\cup I'} = d^*_{S\cup I'}\mathcal{C}d_{R\cup I'}$ where d_X represents the restriction of d to the subspace indexed by X. By the Cauchy-Binet theorem,

$$\det d^*_{S\cup I'}\mathcal{C}d_{R\cup I'} = \sum_Y \det(d^Y_{S\cup I'})^* \det \mathcal{C}^Y_Y \det d^Y_{R\cup I'}, \tag{13}$$

where the sum is over collections of edges Y of cardinality $|S\cup I'| = |R \cup I'|$.

The terms in the sum (13) are collections of edges Y in the graph $\mathcal{G}$. The number of components of the subgraph with edges Y is at least

$$|\text{vertices}| - |\text{edges}| = |R \cup S \cup Q \cup I'| - |Y| = |R| + |Q|.$$

We claim that for each nonzero term in the sum (13), each component is a tree containing exactly one element of $S \cup Q$. If some component did not contain a point of $S \cup Q$, then the function which is 1 on this component and zero elsewhere would be in the kernel of $d^Y_{R\cup I'}$ whose determinant would then be zero. Since the number of components is $|R| + |Q| = |S| + |Q|$ each component must contain exactly one point of $S \cup Q$. Similarly each component also contains exactly one element of $R \cup Q$ (since $\det d^Y_{S\cup I} \neq 0$). Thus each component is a tree containing either a unique $q \in Q$ (and no point of $R \cup S$), or a tree containing both a unique $r \in R$ and a unique $s \in S$. The total number of components is $|R| + |Q|$ so each element of Q, R and S is in a component. In particular there is a permutation ρ and for each i a component joining r_i, the ith element of R, to $s_{\rho(i)}$, the $\rho(i)$th element of S. The weight of a component is the product of its edge weights. The sign of the term corresponding to the subset Y can be determined as follows. Each component connected to Q occurs in both $\det d^Y_{R\cup I'}$ and $\det d^Y_{S\cup I'}$ and so contributes sign $+1$. If we postmultiply $d_{S\cup I'}$ by the permutation matrix P which permutes the rows according to ρ then the sign of the corresponding terms in $Pd_{S\cup I'}$ and $d_{R\cup I'}$ are the same since they corresponding to the same choice of rows. Since $\det P = (-1)^\rho$ this completes the proof. □

4.3.3. *Other groves.* Theorem 4.3 shows that, when divided by the factor Pr(unc) (the probability of the uncrossing) certain grove probabilities are given by noninterlaced minors of L: those groves which correspond to noninterlaced pairings, for example.

Remarkably, one can give a similar expression for *all* grove types: see Theorem 4.4 below.

For a partition τ of $\{1, \ldots, n\}$ (planar or not) we define

$$L_\tau = \sum_F \prod_{\{i,j\} \in F} L_{i,j}, \tag{14}$$

where the sum is over those spanning forests F of the complete graph on n vertices $1, \ldots, n$ for which the trees of F span the parts of τ. For example we have

$$L_{(123|4)} = L_{12}L_{13} + L_{12}L_{23} + L_{13}L_{23}$$

and

$$L_{(13|24)} = L_{13}L_{24}.$$

THEOREM 4.4 ([**19**]). *For any planar partition σ we have*

$$\frac{\Pr(\sigma)}{\Pr(\text{unc})} = \sum_\tau \mathcal{P}_{\sigma,\tau} L_\tau$$

where $\mathcal{P}$ is the integer matrix defined below.

The rows of the matrix $\mathcal{P}$ are indexed by planar partitions, and the columns are indexed by all partitions. In the case of $n = 4$ nodes, the matrix $\mathcal{P}$ is

$$\begin{array}{c|ccccccccccccccc}
 & 1|2|3|4 & 12|3|4 & 13|2|4 & 14|2|3 & 23|1|4 & 24|1|3 & 34|1|2 & 12|34 & 14|23 & 1|234 & 2|134 & 3|124 & 4|123 & 1234 & 13|24 \\
\hline
1|2|3|4 & 1 & 0 & 0 & 0 & 0 & 0 & 0 & 0 & 0 & 0 & 0 & 0 & 0 & 0 & 0 \\
12|3|4 & 0 & 1 & 0 & 0 & 0 & 0 & 0 & 0 & 0 & 0 & 0 & 0 & 0 & 0 & 0 \\
13|2|4 & 0 & 0 & 1 & 0 & 0 & 0 & 0 & 0 & 0 & 0 & 0 & 0 & 0 & 0 & 0 \\
14|2|3 & 0 & 0 & 0 & 1 & 0 & 0 & 0 & 0 & 0 & 0 & 0 & 0 & 0 & 0 & 0 \\
23|1|4 & 0 & 0 & 0 & 0 & 1 & 0 & 0 & 0 & 0 & 0 & 0 & 0 & 0 & 0 & 0 \\
24|1|3 & 0 & 0 & 0 & 0 & 0 & 1 & 0 & 0 & 0 & 0 & 0 & 0 & 0 & 0 & 0 \\
34|1|2 & 0 & 0 & 0 & 0 & 0 & 0 & 1 & 0 & 0 & 0 & 0 & 0 & 0 & 0 & 0 \\
12|34 & 0 & 0 & 0 & 0 & 0 & 0 & 0 & 1 & 0 & 0 & 0 & 0 & 0 & 0 & -1 \\
14|23 & 0 & 0 & 0 & 0 & 0 & 0 & 0 & 0 & 1 & 0 & 0 & 0 & 0 & 0 & -1 \\
1|234 & 0 & 0 & 0 & 0 & 0 & 0 & 0 & 0 & 0 & 1 & 0 & 0 & 0 & 0 & 1 \\
2|134 & 0 & 0 & 0 & 0 & 0 & 0 & 0 & 0 & 0 & 0 & 1 & 0 & 0 & 0 & 1 \\
3|124 & 0 & 0 & 0 & 0 & 0 & 0 & 0 & 0 & 0 & 0 & 0 & 1 & 0 & 0 & 1 \\
4|123 & 0 & 0 & 0 & 0 & 0 & 0 & 0 & 0 & 0 & 0 & 0 & 0 & 1 & 0 & 1 \\
1234 & 0 & 0 & 0 & 0 & 0 & 0 & 0 & 0 & 0 & 0 & 0 & 0 & 0 & 1 & 0.
\end{array}$$

For example, the row for $1|234$ tells us

$$\frac{\Pr(1|234)}{\Pr(1|2|3|4)} = L_{(1|234)} + L_{(13|24)} = L_{2,3}L_{3,4} + L_{2,3}L_{2,4} + L_{2,4}L_{3,4} + L_{1,3}L_{2,4}$$

and the row for 12|34 tells us

$$\frac{\Pr(12|34)}{\Pr(1|2|3|4)} = L_{(12|34)} - L_{(13|24)} = L_{1,2}L_{3,4} - L_{1,3}L_{2,4}.$$

We call this matrix $\mathcal{P}$ the **projection matrix from partitions to planar partitions**, since it can be interpreted as a map from the vector space whose basis vectors are indexed by all partitions to the vector space whose basis vectors are indexed by planar partitions, and the map is the identity on planar partitions. For example, the column for 13|24 tells us

$$13|24 \text{ projects to } \quad -12|34 - 14|23 + 1|234 + 2|134 + 3|124 + 4|123.$$

The projection matrix may be computed using some simple combinatorial transformations of partitions. Given a partition τ, the τth column of $\mathcal{P}$ may be computed by repeated application of the following transformation rule, until the resulting formal linear combination of partitions only involves planar partitions. The rule is a generalization of the transformation

$$13|24 \to -12|34 - 14|23 + 1|234 + 2|134 + 3|124 + 4|123$$

which can be represented as a sort of skein diagram:

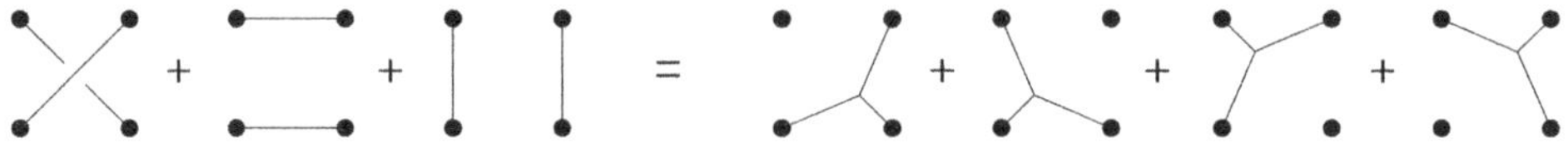

We generalize this rule to partitions τ containing additional items and parts as follows. If partition τ is nonplanar, then there will exist items $a < b < c < d$ such that a and c belong to one part, and b and d belong to another part. Arbitrarily subdivide the part containing a and c into two sets A and C such that $a \in A$ and $c \in C$, and similarly subdivide the part containing b and d into $B \ni b$ and $D \ni d$. Let the remaining parts of partition τ (if any) be denoted by "rest." Then the transformation rule is

$$\text{(Rule 1)} \quad AC|BD|\,\text{rest} \to A|BCD|\,\text{rest} + B|ACD|\,\text{rest} + C|ABD|\,\text{rest} + D|ABC|\,\text{rest} - AB|CD|\,\text{rest} - AD|BC|\,\text{rest}\,.$$

THEOREM 4.5 ([**19**]). *Any partition τ may be transformed into a formal linear combination of planar partitions by repeated application of Rule 1, and the resulting linear combination does not depend on the choices made when applying Rule 1, so that we may write*

$$\tau \to \sum_{\text{planar partitions } \sigma} \mathcal{P}_{\sigma,\tau}\sigma.$$

4.4. Minimality and electrical equivalence.

4.4.1. *Medial graph.* If $\mathcal{G}$ is a circular planar network or network on a surface, the **medial graph** of $\mathcal{G}$ is the graph with a vertex $v(e)$ for every edge e of $\mathcal{G}$, and an edge connecting $v(e)$ and $v(e')$ if e and e' share a vertex and are adjacent edges (in cyclic order) out of this vertex. The medial graph is regular of degree 4. We usually cut the edges of the medial graph which separate the nodes from ∞, so that the medial graph has two half-edges (called "**stubs**") adjacent to each node, see Figure 3.

A **strand** of the medial graph M is a maximal path in the medial graph which continues "straight" at each vertex of M, that is, turns neither left nor right. There is a strand starting at each stub and ending at another stub, and possibly other strands which form closed loops.

If $\mathcal{G}$ is drawn in a disk, with the nodes on the bounding circle, the strands of a medial graph are the boundaries of a cell decomposition of the disk. The cells come in two types, corresponding to vertices and faces of $\mathcal{G}$ (between two adjacent nodes there is a cell corresponding to an exterior face of $\mathcal{G}$).

A nice property of medial strands is that, for minimal graphs (see the next section), they are boundaries of the support of equilibrium potentials. Let us explain.

LEMMA 4.6. **[6]** *Suppose $\mathcal{G}$ is minimal. Let a, b be stubs connected by a medial strand. This medial strand divides the disk into two parts U_+, U_-. Then there is an equilibrium potential f (a function harmonic on I and with Dirichlet boundary conditions on $\mathcal{N}$), not identically zero, with harmonic conjugate f^* with the property that both f and f^* are identically zero on (cells in) U_-.*

For an example see Figure 4. Furthermore we have

LEMMA 4.7. **[6]** *If G is minimal and a, b are two stubs, let C_+ and C_- be the two arcs of the boundary of the disk separated by a and b. Then a and b are connected by a medial strand in $\mathcal{G}$ if and only if there are equilibrium potentials f_+ and f_- and conjugates respectively g_+, g_- such that f_+, g_+ are zero on nodes and exterior faces of C_+, and f_-, g_- are zero on nodes and exterior faces of C_-.*

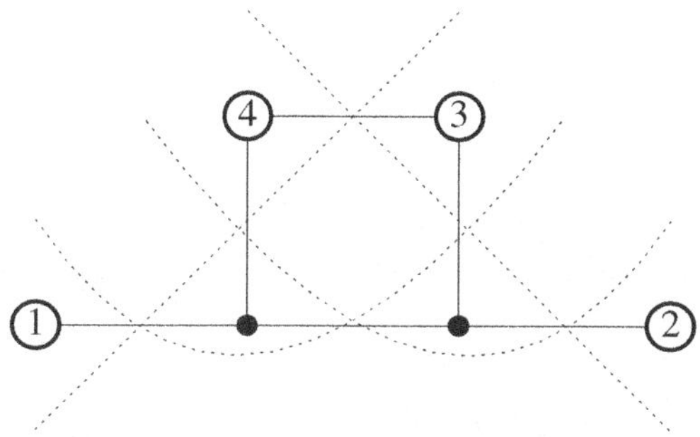

FIGURE 3. The medial graph (dotted) for a 4-node network (solid).

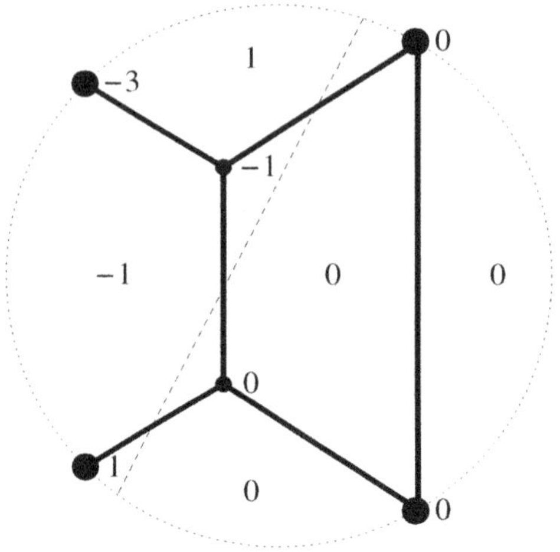

FIGURE 4. A network (solid) with a strand (dashed) and an equilibrium potential and dual which are zero on one side. In this example all conductances are 1.

4.4.2. *Minimality and electrical transformations.* A circular planar network is said to be **minimal** if the medial strands have the properties

(1) There are no closed loops.
(2) A strand does not intersect itself.
(3) No two strands intersect more than once.

The reason we define minimal graphs is that every circular planar network is equivalent, in the sense of having the same response matrix, to a minimal circular planar network. Moreover minimal graphs are precisely those in which the reconstruction has a unique solution [**6**].

In [**7**] it was shown that any circular planar network can be converted into a minimal circular planar network by a sequence of local transformations called electrical transformations. An **electrical transformation** is a local rearrangement of the graph and conductances of one of the types shown in Figure 5. They consist of:

(1) Removing a "dead branch" (remove a non-node vertex of degree 1 and the edge connecting it to the rest of the graph) or a self-loop.
(2) Replacing two conductors c_1, c_2 in series with a single conductor of conductance $\frac{c_1c_2}{c_1+c_2}$ (as long as the common vertex is not a node).
(3) Replacing parallel edges of conductances c_1, c_2 with a single edge of conductance $c_1 + c_2$.
(4) the "Y-Delta" move, also called star-triangle move: an internal vertex of degree three with edges of conductance c_1, c_2, c_3 is replaced by a triangle with edges of conductance $\frac{c_1c_2}{c_1+c_2+c_3}, \frac{c_1c_3}{c_1+c_2+c_3}, \frac{c_2c_3}{c_1+c_2+c_3}$ as in the diagram.

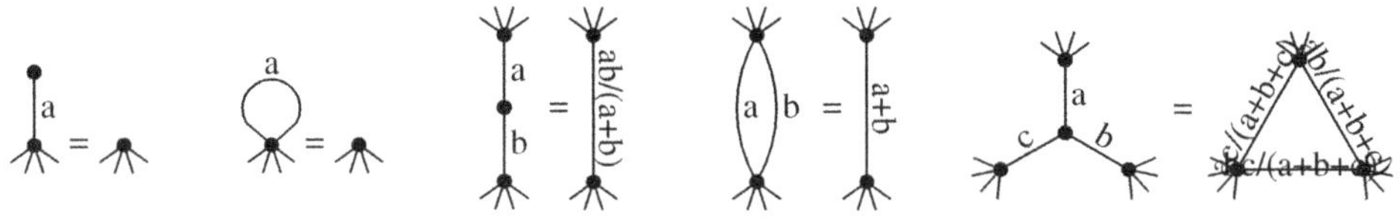

FIGURE 5. The electrical transformations.

The reverse of a Y-Delta move is a Delta-Y move and is also allowed. The inverse transformation takes conductances c_1, c_2, c_3 on the "Delta" to conductances

$$\frac{c_1c_2 + c_1c_3 + c_2c_3}{c_1}, \frac{c_1c_2 + c_1c_3 + c_2c_3}{c_2}, \frac{c_1c_2 + c_1c_3 + c_2c_3}{c_3}$$

on the Y. For simplicity when we say "Y-Delta move" we mean either a Y-Delta or its reverse.

We say that circular planar networks $\mathcal{G}, \mathcal{G}'$ on the same number of nodes are **topologically equivalent** if one can be converted to the other as graphs (that is, ignoring conductances), using electrical transformations. We define circular planar networks $\mathcal{G}, \mathcal{G}'$ to be **electrically equivalent** if they have the same response matrix.

4.4.3. *Topological equivalence.* It is not hard to show that any circular planar network is topologically equivalent to a minimal circular planar network, that is, can be converted to a minimal network by electrical transformations. The idea is to simply chart how the moves act on medial strands: Each of the first three types of moves decreases the number of crossings of the medial strands (that is, the number of edges of the network). We isotope the strands, leaving their endpoints fixed, until each strand has no self-intersections and crosses each other strand at most once. During this isotopy, each time a strand crosses an intersection of other strands we perform a Y-Delta move on the graph; other singularities encountered will decrease the number of crossings and so are finite in number.

This argument shows that for the purposes of determining topological equivalence one can assume $\mathcal{G}$ and $\mathcal{G}'$ are minimal. What is a bit harder to show is that the "minimization" process leads to a unique minimal graph, up to Y-Delta moves. That is, if $\mathcal{G}'$ and $\mathcal{G}''$ are minimal graphs obtained from the same graph $\mathcal{G}$ then $\mathcal{G}'$ can be converted to $\mathcal{G}''$ using only Y-Delta moves (and Delta-Y moves). See Theorem 4.8 below.

For a given minimal circular planar graph $\mathcal{G}$ with n nodes, the medial graph has $2n$ stubs, one starting to the left and right of each node. The medial paths connect these stubs in pairs. Let π denote the pairing of the stubs by the medial paths; if we label the stubs from 1 to $2n$ in cclw order then we can think of π as a fixed-point free involution of $\{1, 2, \ldots, 2n\}$.

THEOREM 4.8 ([**6**]). *Two minimal graphs $\mathcal{G}, \mathcal{G}'$ are topologically equivalent if and only if they have the same medial strand stub involution.*

PROOF. If two minimal graphs have the same medial strand involution then one can isotope the strands without forming any new crossings so the two graphs are isomorphic: for example put the nodes and stubs in general position on a circle respecting their circular order; now isotope the strands to straight chords across the disk.

Conversely, by Lemma 4.7 the response matrix of a minimal graph $\mathcal{G}$ determines its stub involution, since the supports of equilibrium potentials and their harmonic conjugates are determined by L. □

4.4.4. *Well-connected and non-well-connected networks.* Recall that a network is well-connected if all non-interlaced minors of L are strictly positive. Equivalently, for any non-interlaced subsets of nodes A, B, there is a grove connecting points of A to those of B in pairs. (If $|A| = |B| = k$ one can just look for a set of k pairwise disjoint paths connecting points of A to those of B; such a set is easily completed to a grove by adding extra edges.)

What stub involution corresponds to a well-connected network? It is not hard to construct well-connected graphs. One nice family is illustrated in Figure 6. We leave it to the reader to show that these graphs are well-connected, and have stub involution π_{well} which pairs i with $n+i$ for each i. That is, each stub is paired with the diametrically opposite stub.

COROLLARY 4.9. *For a minimal graph $\mathcal{G}$ on n nodes, the following are equivalent:*

(1) *$\mathcal{G}$ is well-connected.*
(2) *Its stub involution is π_{well}.*
(3) *$\mathcal{G}$ is topologically equivalent to Γ_n.*

Another nice family of well-connected graphs which are in addition circularly symmetric (for n odd) and nearly circularly symmetric (for n even) is given in Figure 7. See Figure 8 for a third family.

What about non-well connected graphs? Is there any structure to the set of topological equivalence classes of n-node networks? See below.

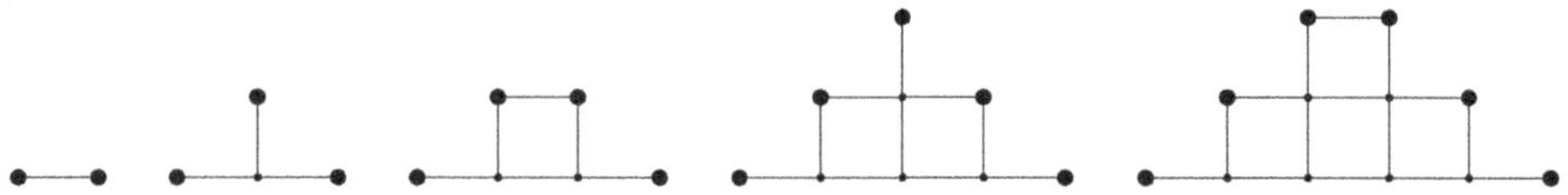

FIGURE 6. The standard networks Γ_k for $k = 2, \ldots, 6$. These are well-connected.

FIGURE 7. Some (almost) circularly symmetric well-connected graphs.

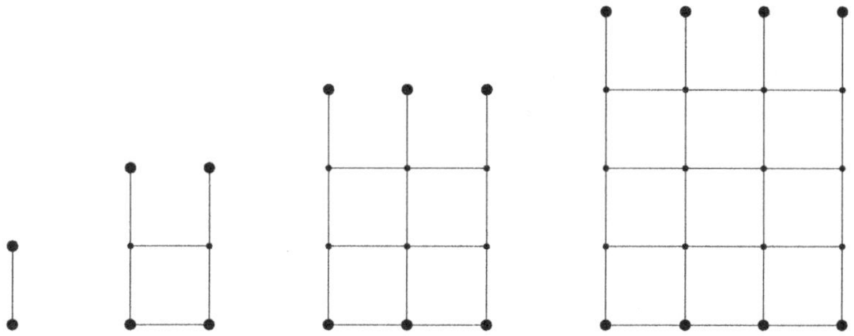

FIGURE 8. A different family of well-connected graphs for even n.

4.4.5. *Electrical equivalence.* We discussed topological equivalence above. Regarding electrical equivalence, we have

THEOREM 4.10. *Electrical transformations do not change the response matrix.*

PROOF. It suffices to show that electrical transformations preserve current flow. More precisely, the current flow on the "before" network will be equal to the current flow on the "after" network if defined appropriately on any added edges.

This is clear for transformations of type 1 (removing a dead branch or self-loop) since these edges have no current flow. For a type 3 move (parallel reduction) the current flow on c_1 is $c_1 df(e)$ where f is the corresponding potential on the vertices. Similarly the current flow across c_2 is $c_2 df(e)$. The current flow on the combined edge is $(c_1 + c_2)df(e)$ which is the sum of the current flows on the individual edges.

For a type 2 (series) transformation, the relevant equation is $c_1 df(e_1) = c_2 df(e_2)$ since no current is lost at the center vertex. This implies that

$$c_1 df(e_1) = \frac{c_1 c_2}{c_1 + c_2}(df(e_1) + df(e_2)),$$

and $df(e_1) + df(e_2)$ is the potential drop across the combined edges.

A similar computation works for the Y-Delta move. When converting from a Y to a Delta, the current along the edges of the Y directed towards the central vertex is $c_1 df_1, c_2 df_2, c_3 df_3$ (where the df_i are the potential drops) which must add to zero since no current is lost at the central vertex:

$$c_1 df_1 + c_2 df_2 + c_3 df_3 = 0.$$

A short computation then shows that

$$\begin{aligned} c_1 df_1 &= C_{12}(df_1 - df_2) + C_{13}(df_1 - df_3), \\ c_2 df_2 &= C_{12}(df_1 - df_2) + C_{13}(df_1 - df_3), \\ c_3 df_3 &= C_{12}(df_1 - df_2) + C_{13}(df_1 - df_3) \end{aligned}$$

for arbitrary df_1, df_2, df_2 when and only when $C_{ij} = \frac{c_i c_j}{c_1+c_2+c_3}$. □

As a consequence of this theorem, for determining the space of response matrices of planar networks we need only consider minimal networks.

4.5. Reconstruction and the space of networks. Given a matrix which satisfies the conditions and inequalities of Theorem 4.1, is it the response matrix of a circular planar network? How can one determine the conductances from L? This is the **Reconstruction Problem** or **EIT problem**. It is closely related to the **Electrical Equivalence Problem**: what sets of minimal networks are electrically equivalent?

4.5.1. *Algorithm.* Both [**8**] and [**6**] gave an iterative algorithm for reconstruction on any minimal graph. The idea is as follows. Any network is equivalent to one which has either an edge connecting adjacent nodes or a node of degree 1 (take a strand ab with the property that there are no strands completely contained on one side U_+ of it; using Y-Delta moves, move all crossings in U_+ of other strands to the other side U_-. The node adjacent to a and inside U_+ will either have degree 1 or be connected to its adjacent node on the other side of a).

One can now compute the conductance on this edge e from L: find the potential with values zero on side U_+ of strand ab. Then the conductance c_e is easily obtained from the current.

Removing the edge e (if it connected nodes) or contracting the edge e (if it is adjacent to a degree-1 node) results in a new network which is still minimal, and the new response matrix is a simple rational function of the old response matrix.

This algorithm shows that the conductances are *rational functions of the* L_{ij}. However it is not so easy to use this to give an explicit formula for the conductances as a function of the response matrix. In [**20**] an explicit formula is given for the reconstruction problem for the standard graphs Γ_n of Figure 6, which expresses c_e as a ratio of polynomials in the L_{ij} which are Pfaffians of skew-symmetric matrices formed from L. This can be considered in some sense the best possible result for Γ_n since the polynomials of the L_{ij} are shown there to be irreducible in general. However this does not preclude the possibility that for other well-connected networks there is a simpler formula: see [**21**].

4.5.2. *Partial order on networks.* By allowing some conductances to go to zero and others to tend to ∞ a network degenerates (topologically, we remove edges of conductance zero and contract edges of conductance ∞).

There is a partial order on networks under this degeneration. We define $\mathcal{G} \leq \mathcal{G}'$ if a network equivalent to $\mathcal{G}$ can be obtained from a network equivalent to $\mathcal{G}'$ by contraction and deletion of edges.

One can show that every minimal network is a minor (in the sense of contraction and/or deletion of edges) of a well-connected network. So the well-connected network is the unique maximal element in this partial order. If we allow disconnected networks, the networks which are minimal are those whose stub involution is a non-crossing matching; see Figure 10.

LEMMA 4.11. *If $\mathcal{G}, \mathcal{G}'$ are minimal circular planar networks and $\mathcal{G} \leq \mathcal{G}'$ then there is a sequence $\mathcal{G}_1 \leq \mathcal{G}_2 \leq \cdots \leq \mathcal{G}_k = \mathcal{G}'$ of minimal networks where $\mathcal{G}_1$ is equivalent to $\mathcal{G}$ and $\mathcal{G}_i$ is obtained from a network equivalent to $\mathcal{G}_{i+1}$ by deletion or contraction of a single edge.*

PROOF. First suppose that $\mathcal{G}$ is obtained from $\mathcal{G}'$ by deleting a single edge e (the case of contraction is equivalent by duality). Here is how one minimizes the resulting network: let a, b, c, d in cclw order be the stubs of the strands ac and bd crossing at edge e. Let U_1, U_2, U_3, U_4 be the four arcs of the circle separated by a, b, c, d, so that U_1 is the arc from a to b, etc. Using Y-Delta moves we can first comb strands so that strands connecting stubs from U_1 to U_3 cross "left" of e, that is, do not enter the region delimited by the path bec, and strands connecting stubs from U_2 and U_4 cross "below" e, that is, do not enter the region delimited by the path ced. See the Figure 9. Now we resolve the crossing at e, creating a strand connecting a and b and another connecting c and d. The strand ab forms a bigon with any strand from U_2 to U_4. To minimize the resulting network we resolve all crossings of these strands with the semi-strand from b to e as indicated in Figure 9. The resulting network is now minimized (and any minimization results in a network equivalent to this one). If we perform this sequence of resolutions in order of the crossings along the semistrand be from the boundary towards the center, (resolving the crossing at e last) the intermediate networks are all minimal. This completes the proof in the case when $\mathcal{G}$ is obtained from $\mathcal{G}'$ by resolution of a single crossing.

In the general case, we contract and delete multiple edges of $\mathcal{G}'$, then minimize to get $\mathcal{G}$. Take a path in the space of conductances which sends the desired conductances to zero or infinity, one after another, in any order. Let $L(t)$ be the corresponding path in the space of response matrices. When we

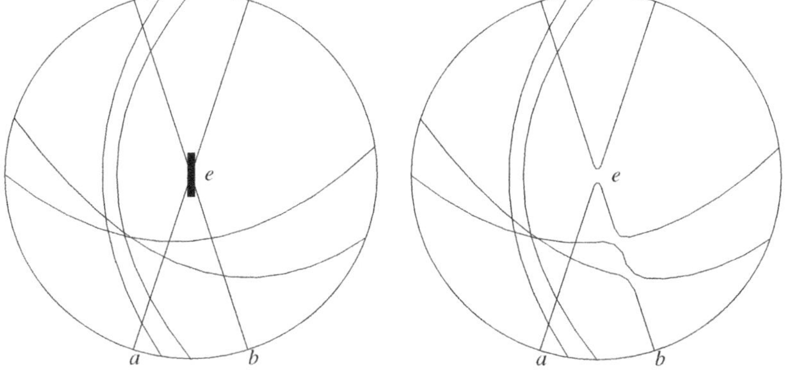

FIGURE 9. When a conductance e goes to zero, comb strands left and down (left figure); then resolve strand crossings between b and e (right figure).

minimize the resolution of a single crossing as above, the resulting L matrix does not change (in the sense that no additional minors become zero), since minimization leaves L invariant. Thus the sequence of resolutions is replaced with a sequence of "minimal" resolutions, that is, resolutions which preserve minimality. □

This partial order on networks induces a partial order on fixed-point free involutions of $\{1, \ldots, 2n\}$. This partial order is graded by the number of crossings in the involution, that is, the number of pairs $i < j < k < \ell$ with i paired with k and j paired with ℓ. Moving one level down in the partial order corresponds to resolving exactly one crossing which is adjacent to the boundary (in the sense that there is a path in the disk from the crossing to the boundary circle which does not meet any strand). Algebraically, moving up one level corresponds to the operation of conjugating the involution by a transposition in such a way as to increase the number of crossings by 1.

See Figure 10 for the Hasse diagram of the partial order in the case $n = 3$.

This partial order describes the cell structure of the space of networks.

4.5.3. *Minimal sets of inequalities.* What is the structure of the set of response matrices of all circular planar networks? The set Ω of response matrices of circular planar networks on n nodes is a closed subset of $\mathbb{R}^{n(n-1)/2}$

FIGURE 10. The partial order on circular planar networks with three nodes. Only the medial graphs are shown.

if we use the coordinates $\{L_{ij}\}_{n\geq i>j\geq 1}$. It is a semi-algebraic set, that is, defined by a set of algebraic inequalities $\det L_A^B \geq 0$ for the non-interlaced subsets A, B of nodes. Its interior Ω^+ is defined by using strict inequalities, and corresponds to the well-connected networks.

The number of inequalities defining Ω is exponential in n. However there is a smaller set of $n(n-1)/2$ inequalities which defines Ω^+. See below.

If we consider a topological equivalence class (defined by a stub involution π) of graphs, the response matrices of networks supported on these graphs form a subset $\Omega_\pi \subset \Omega$, which is a subset of the boundary of Ω if $\pi \neq \pi_{\text{well}}$. Ω has the structure of a cell-complex, in which the Ω_π are the cells, see Lam and Pylyavskyy [**23**]. Is each cell of dimension k defined by k minor inequalities (and $n(n-1)/2 - k$ equalities)? The answer is probably yes, but at the moment our understanding is limited. A similar situation where the cell structure is explicitly worked out is Postnikov [**29**] who deals with totally positive/totally nonnegative matrices and the totally nonnegative Grassmannian.

Let us discuss here only the minimal sets of inequalities defining the set Ω^+. Somewhat remarkably, there are many different sets $\binom{n}{2}$ minor determinants of L whose positivity implies the positivity of all non-interlaced minors. The situation resembles that of a cluster algebra but so far we have been unable to find the relevant cluster structure.

One easy-to-remember set of $\binom{n}{2}$ minors are the **central minors**, defined as follows. Suppose first n is odd. Define $M_{i,1}$ for $1 \leq i \leq (n-1)/2$ to be the minor L_A^B with $A = \{1, 2, \ldots, i\}$ and $B = \{(n-1)/2+1, (n-1)/2+2, \ldots, (n-1)/2+i\}$. Let $M_{i,j}$ be the minor obtained from $M_{i,1}$ by rotating the indices cyclically by $j-1$, so that $A = \{j, j+1, \ldots, j+i\}$ and likewise for B.

These minors are called central minors since, if we arrange the n points evenly spaced on a circle, and connect each element of A to the corresponding point of B opposite it, we get a set of i parallel chords which are "central" in the sense that they are as close to a fixed diagonal of the circle as possible while remaining disjoint from each other. See Figure 11 for an example. In the case n is even we modify the above; define $M_{i,j}$ for i even as before, but for i odd we have, for each diagonal, two choices of positioning of the chords to make them "most central"; choose one of these arbitrarily. These "off-center" minors are part of our set of central minors. See Figure 12 for a natural choice.

THEOREM 4.12 ([**21**]). *If the central minors $\{M_{i,j}\}$ are positive then all noninterlaced minors are positive.*

The following proof is lifted directly from [**21**].

PROOF. This follows using two identities for minors.

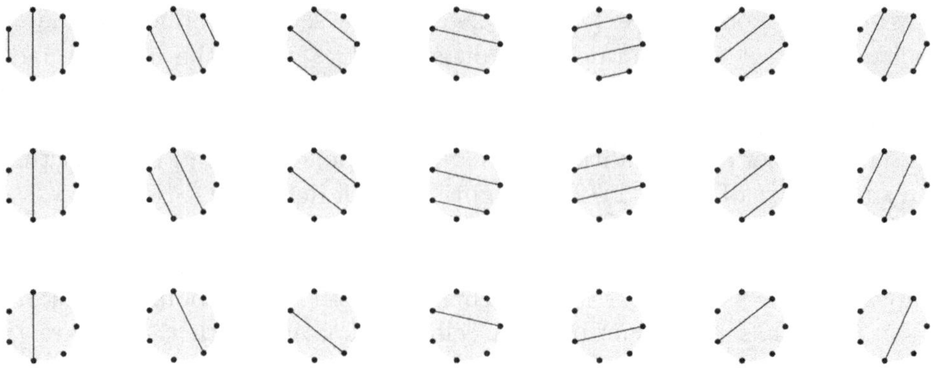

FIGURE 11. The 21 central minors for a well-connected network on 7 nodes.

Let U be a matrix, and let a, b, c index some of its columns, and z, d index some of its rows. Then it is elementary that

$$0 = \begin{vmatrix} U_{z,a} & U_{z,b} & U_{z,c} \\ U_{z,a} & U_{z,b} & U_{z,c} \\ U_{d,a} & U_{d,b} & U_{d,c} \end{vmatrix} = U_{z,a} \begin{vmatrix} U_{z,b} & U_{z,c} \\ U_{d,b} & U_{d,c} \end{vmatrix} - U_{z,b} \begin{vmatrix} U_{z,a} & U_{z,c} \\ U_{d,a} & U_{d,c} \end{vmatrix} + U_{z,c} \begin{vmatrix} U_{z,a} & U_{z,b} \\ U_{d,a} & U_{d,b} \end{vmatrix}.$$

Suppose U is invertible, and let M denote its inverse. Let $r_1, \dots, r_k$ denote the row indices of M (and column indices of U), and $c_1, \dots, c_k$ denote the column indices of M (and row indices of U). Dividing through by $(\det U)^2$ and using Jacobi's identity relating minors of a matrix to minors of its

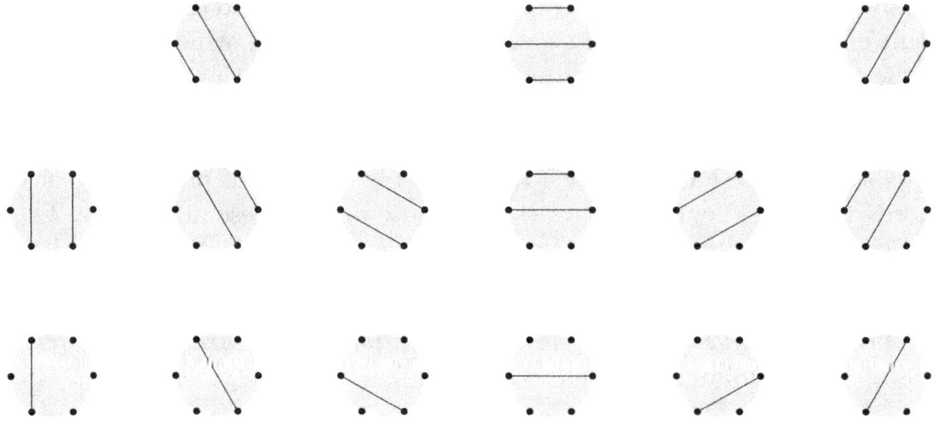

FIGURE 12. The 15 central minors for a well-connected network on 6 nodes.

inverse, we obtain

$$0 = \det(M_{i,j})^{i\neq a}_{j\neq z} \det(M_{i,j})^{i\neq b,c}_{j\neq z,d} - \det(M_{i,j})^{i\neq b}_{j\neq z} \det(M_{i,j})^{i\neq a,c}_{j\neq z,d} + \det(M_{i,j})^{i\neq c}_{j\neq z} \det(M_{i,j})^{i\neq a,b}_{j\neq z,d}.$$

Since the column z is always excluded, we may as well suppose that M is a $k \times (k-1)$ matrix. Then

$$\det(M_{i,j})^{i\neq b}_{j} \det(M_{i,j})^{i\neq a,c}_{j\neq d} = \det(M_{i,j})^{i\neq a}_{j} \det(M_{i,j})^{i\neq b,c}_{j\neq d} + \det(M_{i,j})^{i\neq c}_{j} \det(M_{i,j})^{i\neq a,b}_{j\neq d}.$$

For example, taking $M = L^{9,8,7,6,5}_{1,2,3,4}$ and $a,b,c,d = 9,8,5,4$, we have

$$\left|L^{9,7,6,5}_{1,2,3,4}\right| = \frac{\left|L^{8,7,6,5}_{1,2,3,4}\right| \left|L^{9,7,6}_{1,2,3}\right| + \left|L^{9,8,7,6}_{1,2,3,4}\right| \left|L^{7,6,5}_{1,2,3}\right|}{\left|L^{8,7,6}_{1,2,3}\right|}.$$

We can denote this pictorially as

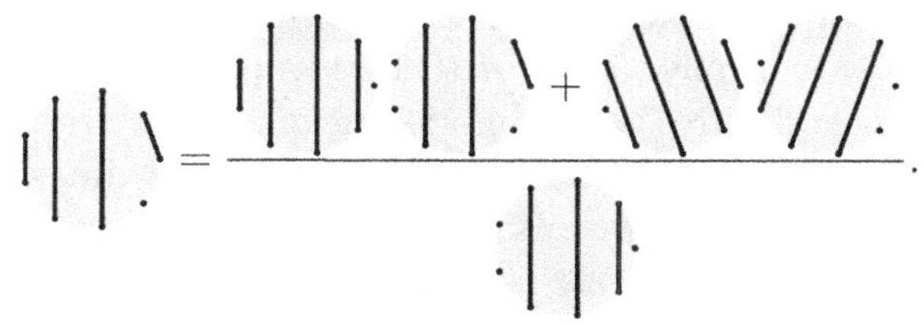

We call transformations of this type the "jaw move".

For a given non-interleaved determinant interspersed with at least one isolated node, we can take b to be one of the interspersed isolated nodes, and a and c to be the first and last of the nodes on the same side as b, and d to be either first or last node on the other side as b. With this choice of a, b, c, d, the jaw move expresses the original determinant as a positive rational function of "simpler" non-interleaved determinants, where a determinant is simpler if it has fewer strands, or else the same number of strands but fewer interspersed isolated nodes.

By repeated application of the jaw rule, any non-interleaved determinant can be expressed as a positive rational function of non-interleaved contiguous determinants.

The other identity that we need is based on Dodgson condensation (and is also called the Desnanot-Jacobi identity). It can be derived in a similar way as the jaw move, with the elementary starting identity

$$U_{c,a}U_{d,b} = \begin{vmatrix} U_{c,a} & U_{c,b} \\ U_{d,a} & U_{d,b} \end{vmatrix} + U_{c,b}U_{d,a}.$$

Dividing through by $(\det U)^2$ and using Jacobi's identity, we obtain

$$\det(M_{i,j})^{i\neq a}_{j\neq c}\det(M_{i,j})^{i\neq b}_{j\neq d} = \det(M_{i,j})^{i}_{j}\det(M_{i,j})^{i\neq a,b}_{j\neq c,d}$$
$$+ \det(M_{i,j})^{i\neq b}_{j\neq c}\det(M_{i,j})^{i\neq a}_{j\neq d}.$$

For example, taking $M = L^{8,7,6}_{1,2,3}$ and $a,b,c,d = 8,6,3,1$, we obtain

$$\left|L^{8,7}_{1,2}\right| = \frac{\left|L^{8,7,6}_{1,2,3}\right|\left|L^{7}_{2}\right| + \left|L^{7,6}_{1,2}\right|\left|L^{8,7}_{2,3}\right|}{\left|L^{7,6}_{2,3}\right|}.$$

which we can denote pictorially as

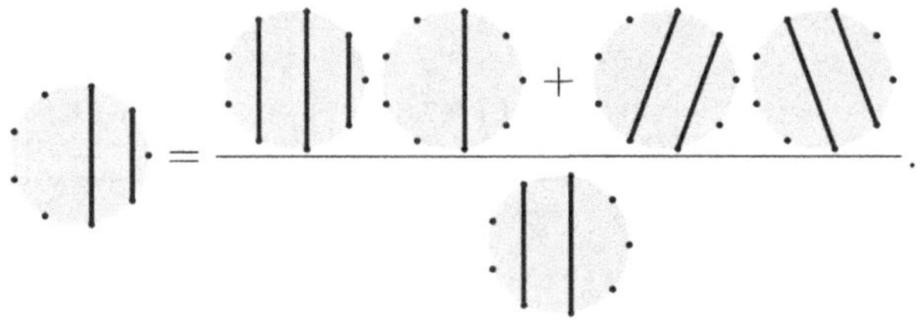

We call transformations of this type the "condensation move".

If both sides of the determinant are contiguous, but the crossings are off-center, then we can pick a and d to be the paired nodes that are most off-center, and b and c to be the nodes next to the crossing but not in the crossing (and closer to the center). Then after a condensation move, the original determinant is expressed as a positive rational function of contiguous crossings that are strictly more central. By repeated application of such condensation moves, any contiguous determination may be expressed as a positive rational function of central contiguous crossings.

The remaining case to check is for n even, with the (minimally) off-center contiguous crossings come in pairs, of which only one is included in the base cluster. Using a condensation move we can express a minimally off-center contiguous crossing in terms of its opposite minimally off-center contiguous crossing and central contiguous crossings as shown below

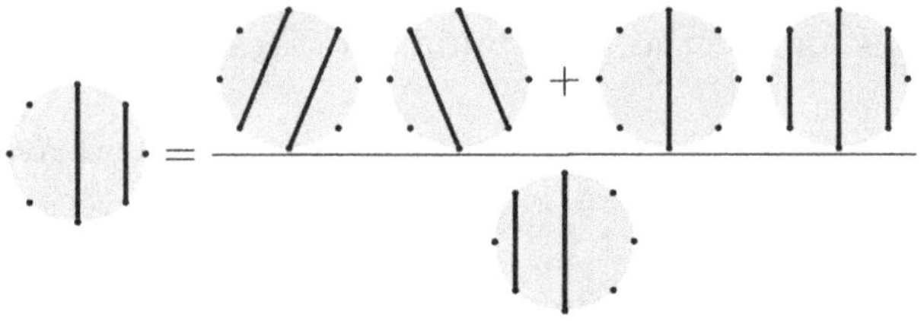

□

4.6. The Jacobian. A remarkable property of the birational map from conductances to the (appropriate set of) L matrix minors for minimal networks is that the Jacobian is ± 1:

THEOREM 4.13 ([**21**]).

$$\det\left(\frac{\partial \log M_i}{\partial \log c_j}\right) = \pm 1.$$

In other words the volume form $\prod_e \frac{dc_e}{c_e}$ on the space of conductances is mapped to ± 1 times the volume form $\prod_i \frac{dM_i}{M_i}$ on the space of L-matrix minors. Note that the matrix entries have a probabilistic interpretation: each M_i can be written $M_i = \Pr(\pi)/\Pr(\mathrm{unc})$ for some noninterlaced pairing π. Thus

$$\frac{\partial \log M_i}{\partial \log c_j} = \frac{c_j}{Z(\pi)}\frac{\partial Z(\pi)}{\partial c_j} - \frac{c_j}{Z_{\mathrm{unc}}}\frac{\partial Z_{\mathrm{unc}}}{\partial c_j}$$

is the difference in the probability that edge e_j is in a random π-pairing, minus the probability that e_j is in a random uncrossing.

The proof of this theorem relies of techniques developed in [**29**] and [**30**] and is too long to give here. Let us just give an example, for a well-connected network on 3 nodes, the Y of Figure 1. The central minors in terms of the conductances are $L_{ij} = \frac{c_i c_j}{c_1+c_2+c_3}$. The above Jacobian matrix (with columns in the order L_{12}, L_{13}, L_{23}) is

$$\begin{pmatrix} 1-\dfrac{1}{c_1+c_2+c_3} & 1-\dfrac{1}{c_1+c_2+c_3} & -\dfrac{1}{c_1+c_2+c_3} \\ 1-\dfrac{1}{c_1+c_2+c_3} & -\dfrac{1}{c_1+c_2+c_3} & 1-\dfrac{1}{c_1+c_2+c_3} \\ -\dfrac{1}{c_1+c_2+c_3} & 1-\dfrac{1}{c_1+c_2+c_3} & 1-\dfrac{1}{c_1+c_2+c_3} \end{pmatrix}$$

whose determinant is 1.

5. Networks on surfaces with nontrivial topology

Some of the material in the preceding section on circular planar networks has been extended to the case of annular networks and more generally networks on surfaces. However the theory is not as complete at present as in the case of circular planar graphs. Moreover beyond the annulus and torus the theory is for the most part nonexistent.

We can nonetheless ask the same questions as arose in the planar case:

(1) What is the natural notion of response matrix?
(2) For which networks can we reconstruct the conductances from the response matrix?
(3) What is the structure of the space of response matrices?
(4) What natural combinatorial objects are relevant?

Our answers, briefly, are as follows. We elaborate on these in the following sections.

(1) We use (the Schur reduction of) the bundle Laplacian for a flat $\mathbb{C}^*$-connection or a flat $\mathrm{SL}_2(\mathbb{C})$-connection on the network.

(2) We can conjecturally reconstruct conductances for minimal networks on the annulus. For other surfaces reconstruction is not in general possible, unless we are given more data: we (conjecturally) need to provide "spectral" data as well.
(3) We don't know what the structure of the space of response matrices is, even for the annulus. However we have a description in the case when there are no nodes, for the annulus and torus.
(4) The natural combinatorial objects are "cycle-rooted spanning forests": forests in which each component has a unique (topologically nontrivial) cycle.

5.1. Vector bundle Laplacian. In the case of a network on a surface it is natural to consider not the standard Laplacian but a more general Laplacian operator, the vector bundle Laplacian, which depends on a connection on the bundle (see definitions below). This is because the standard Laplacian does not contain very much information about the topology of the surface on which the network is embedded...in particular the reconstruction problem is not solvable in general (as an example consider the network on an annulus with two vertices, both nodes, and two edges joining them with different homotopy classes). From the bundle Laplacian we construct a response matrix which contains more information than the one constructed from the standard Laplacian, and conjecturally allows reconstruction for minimal networks.

Let us first define vector bundles and connections on a network.

5.1.1. *Vector bundles and connections.* Given a fixed vector space W, a **W-bundle**, or simply a **vector bundle** on a network $\mathcal{G}$ is the choice of a vector space W_v isomorphic to W for every vertex v of $\mathcal{G}$. A vector bundle can be identified with the vector space $W_{\mathcal{G}} := \oplus_v W_v \cong W^{|\mathcal{G}|}$, called the **total space** of the bundle. A **section** of a vector bundle is an element of $W_{\mathcal{G}}$.

A **connection** Φ on a W-bundle is the choice for each edge $e = vv'$ of $\mathcal{G}$ of an isomorphism $\phi_{vv'}$ between the corresponding vector spaces $\phi_{vv'} : W_v \to W_{v'}$, with the property that $\phi_{vv'} = \phi_{v'v}^{-1}$. This isomorphism is called the **parallel transport** of vectors in W_v to vectors in $W_{v'}$. Two connections Φ, Φ' are said to be **gauge equivalent** if there is for each vertex an isomorphism $\psi_v : W_v \to W_v$ such that the diagram

$$\begin{array}{ccc} W_v & \xrightarrow{\phi_{vv'}} & W_{v'} \\ \downarrow{\psi_v} & & \downarrow{\psi_{v'}} \\ W_v & \xrightarrow{\phi'_{vv'}} & W_{v'} \end{array}$$

commutes. In other words Φ' is just a base change of Φ. Note that the connection has nothing to do with the conductances.

Given an oriented cycle γ in $\mathcal{G}$ starting at v, the **monodromy** of the connection around γ is the element of $\mathrm{End}(W_v)$ which is the product of the parallel transports around γ. Monodromies starting at different vertices on γ are conjugate, as are monodromies of gauge-equivalent connections.

A **line bundle** is a W-bundle where $W \cong \mathbb{C}$, the 1-dimensional complex vector space. In this case given a connection if we choose a basis for each $\mathbb{C}$ then the parallel transport along an edge is just multiplication by an element of $\mathbb{C}^* = \mathbb{C}\backslash\{0\}$. The monodromy of a cycle is in $\mathbb{C}^*$ and does not depend on the starting vertex of the cycle (or gauge).

5.1.2. *Flat connections.* A connection on a network embedded on a surface Σ is said to be a **flat connection** if the monodromy around faces of $\mathcal{G}$ is trivial. This implies that the monodromy is trivial around loops on $\mathcal{G}$ which are null-homotopic as loops on Σ.

It is not hard to see that flat connections, modulo gauge equivalence, are in bijection with homomorphisms of $\pi_1(\Sigma)$ into the group of automorphisms of W.

5.1.3. *The Laplacian.* The Laplacian Δ on a W-bundle with connection Φ is the linear operator $\Delta : W_{\mathcal{G}} \to W_{\mathcal{G}}$ defined by

$$\Delta f(v) = \sum_{v' \sim v} c_{vv'}(f(v) - \phi_{v'v} f(v'))$$

where the sum is over neighbors v' of v.

Note that if the vector bundle is **trivial**, in the sense that $\phi_{vv'}$ is the identity for all edges, this is our the notion of graph Laplacian from (1) (or more precisely, the direct sum of $\dim W$ copies of the Laplacian).

As in the case of the standard network Laplacian we often think of Δ as a matrix $\Delta = (\Delta_{vv'})_{v,v' \in V}$ whose entries are $-c_{vv'}$ times the parallel transport from v' to v. In particular the conductance is acting as a scalar multiplication in W.

Here is an example. Let $\mathcal{G} = K_3$ the triangle with vertices $\{v_1, v_2, v_3\}$. Let Φ be the line bundle connection with $\phi_{v_i v_j} = z_{ij} \in \mathbb{C}^*$. Then in the natural basis, Δ has matrix

$$\Delta = \begin{pmatrix} c_{12} + c_{13} & -c_{12}z_{12} & -c_{13}z_{13} \\ -c_{12}z_{12}^{-1} & c_{12} + c_{23} & -c_{23}z_{23} \\ -c_{13}z_{13}^{-1} & -c_{23}z_{23}^{-1} & c_{13} + c_{23} \end{pmatrix}. \tag{15}$$

5.1.4. *Edge bundle.* One can extend the definition of a vector bundle to the edges of $\mathcal{G}$. In this case there is a vector space $W_e \cong W$ for each edge e as well as each vertex. One defines connection isomorphisms $\phi_{ve} = \phi_{ev}^{-1}$ for a vertex v and edge e containing that vertex, in such a way that if $e = vv'$ then $\phi_{vv'} = \phi_{ev'} \circ \phi_{ve}$, where $\phi_{vv'}$ is the connection on the vertex bundle.

The vertex/edge bundle can be identified with $W^{|\mathcal{G}|+|E|} = W_{\mathcal{G}} \oplus W_E$, where W_E is the direct sum of the edge vector spaces.

A **1-form** (or cochain) is a function on oriented edges which is antisymmetric under changing orientation. If we fix an orientation for each edge, a 1-form is a section of the edge bundle, that is, an element of $W^{|E|}$. We denote by $\Lambda^1(\mathcal{G},\Phi)$ the space of 1-forms and $\Lambda^0(\mathcal{G},\Phi)$ the space of 0-forms, that is, sections of the vertex bundle.

We define a map $d:\Lambda^0(\mathcal{G},\Phi)\to\Lambda^1(\mathcal{G},\Phi)$ by $df(e)=\phi_{ye}f(y)-\phi_{xe}f(x)$ where $e=xy$ is an oriented edge from vertex x to vertex y. We also define an operator $d^*:\Lambda^1\to\Lambda^0$ as follows:

$$d^*\omega(v)=\sum_{e=v'v}\phi_{ev}\omega(e)$$

where the sum is over edges containing v and oriented towards v. Despite the notation, this operator d^* is not a standard adjoint of d unless ϕ_{ev} and ϕ_{ve} are adjoints themselves, that is, if parallel transports are unitary operators (see below).

The Laplacian Δ on Λ^0 can then be defined as the operator $\Delta=d^*\mathcal{C}d$ as before:

$$\begin{aligned}d^*\mathcal{C}df(v)&=\sum_{e=v'v}c_e\phi_{ev}df(e)\\&=\sum_{e=v'v}c_e\phi_{ev}(\phi_{ve}f(v)-\phi_{v'e}f(v'))\\&=\sum_{v'}c_{vv'}(f(v)-\phi_{v'v}f(v'))\\&=\Delta f(v).\end{aligned}$$

We can see from the example (15) above on K_3 that Δ is not necessarily self-adjoint. However if $\phi_{vv'}$ is unitary: $\phi_{vv'}^{-1}=\phi_{v'v}^*$ then d^* will be the adjoint of d for the standard Hermitian inner products on $W^{|\mathcal{G}|}$ and $W^{|E|}$, and so in this case Δ is a Hermitian, positive semidefinite operator. In particular on a line bundle if $|\phi_{vv'}|=1$ for all edges $e=vv'$ then Δ is Hermitian and positive semidefinite.

5.2. Minimal networks. Here we show that any network on a surface can be "minimized", that is, made into a minimal network.

While reconstruction is not possible in general even for minimal networks (see Theorem 7.3 below), the case of the torus teaches us that the map from conductances to the reponse matrix has potentially interesting preimage.

A network $\mathcal{G}$ on a surface Σ is **minimal** if, when lift to the universal cover of Σ, the network is minimal, that is, the lifts of the medial strands do not self-intersect and no two lifts intersect more than once.

As an example, see Figure 20 below.

THEOREM 5.1. *Every network is topologically equivalent to a minimal network.*

PROOF. Any surface with a boundary (or closed surface of genus ≥ 2) has a hyperbolic metric (that is, a Riemannian metric of constant curvature -1) since its universal cover is the Poincaré disk.

Using this metric we can isotope the strands, fixing the endpoints, using a curve-shortening flow until all strands become hyperbolic geodesics. Through this isotopy each singularity encountered either reduces the number of crossings or is a triple crossing, which corresponds to a Y-Delta move of the network. In the end strands which touch the boundary will have the property that, when lifted to $\tilde{\Sigma}$, they meet other strands at most once. More generally, strands in different homotopy classes meet each other at most once in the universal cover.

If two or more intersecting strands with no endpoints (that is, which are closed loops on the surface) have the same homotopy class then we must be more careful since their geodesic representatives will be identical. The situation in the universal cover will resemble that of Figure 13, left, that is, we have a packet of strands which lie near the same geodesic, and other strands cross this packet transversely and exactly once each.

It remains to prove that in this situation we can replace the self-intersecting packet with a "combed" version of itself, where all strands are parallel and non-intersecting (they will still intersect the transversal strands). For this it suffices to consider the situation on an annulus in which the packet strands wind around the annulus forming closed loops, and a certain number of other strands cross this packet transversely. On the universal cover of the annulus, which is a bi-infinite strip, the situation is as on the left in Figure 13.

We consider, then, a periodic graph on a strip $\mathbb{R}\times[0,1]$ with conductances invariant under $x\mapsto x+1$. Suppose we now take n large and truncate the strip, by removing everything far to the left of the origin, say left of $-n$, leaving free boundary conditions there, and removing everything to the right of $+n$, and similarly leaving free boundary conditions there. What remains is a planar graph on a rectangle $[-n,n]\times[0,1]$, and for this planar graph we can "comb out" the packet, as in Figure 13, from the left to the right, so that the strands of the packet become parallel paths, nonintersecting except just left of the right endpoint, where they may cross.

We claim that on the resulting combed finite planar graph, the conductances are nearly periodic, that is, a periodic function plus an error tending to zero as the distance to $\pm n$ increases. To see this we use the interpretation of the response matrix entries as ratios of weighted sums of spanning forests. In [**24**] it is proved that for fixed i,j the truncation only changes $L_{i,j}$ by amounts tending to zero as $n\to\infty$. It is also shown that $L_{i,j}\to 0$ as the distance from i to j increases.

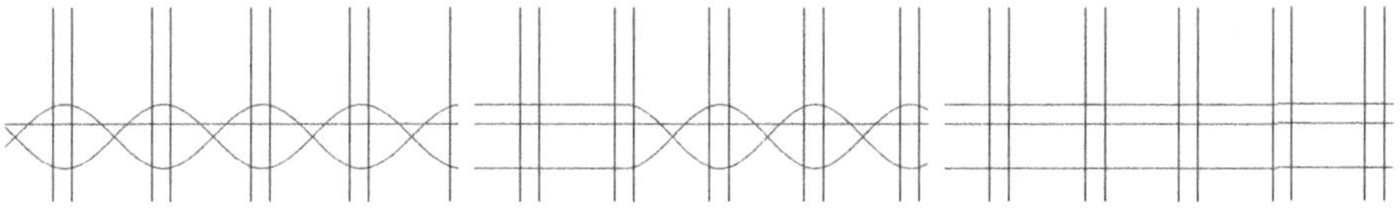

FIGURE 13. Combing a river with three strands.

Take node i on the top and j on the bottom boundary of the strip. The numerator of the entry $L_{i,j}$ is the weighted sum of spanning trees wired at the boundary and with a component connecting i and j (see Theorem 4.3). We claim that for a typical tree the path from i to j does not get near $\pm n$ with high probability, that is, this path does not get far from i or j. If the path extended far beyond j, then (letting $j+k$ denote the translate of j by k) for large k $L_{i,j+k}$ would be of the same order as $L_{i,j}$, since a local change of the long path from i to j would result in a path from i to $j+k$. This contradicts the fact that $L_{i,j+k}$ tends to zero for large k.

Thus $L_{i,j}$ is a function depending less and less (as $n \to \infty$) on the conductances far from i and j. This proves convergence of the conductances as $n \to \infty$, and the translation invariance of $L_{i,j}$ implies the translation invariance of the limiting conductances. □

5.3. Cycle-rooted spanning forests. Where as the standard Laplacian determinant counts spanning trees, in the case of a one- or two-dimensional vector bundle with connection the bundle Laplacian determinant counts objects called cycle-rooted spanning forests. A **cycle-rooted spanning forest** (CRSF) is a subset S of the edges of $\mathcal{G}$, with the property that each connected component has exactly as many vertices as edges (and so has a unique cycle). See Figure 14.

THEOREM 5.2 ([**11**]). *For a line bundle on a finite graph,*

$$\det \Delta = \sum_{CRSFs T} \prod_{e \in T} c_e \prod_{cycles\ of\ T} (2 - w - 1/w)$$

FIGURE 14. A CRSF on a square grid.

where the sum is over all CRSFs T, the first product is over the edges of T, and the second product is over the cycles of T, where $w, 1/w$ are the monodromies of the two orientations of the cycle.

Notice that for a line bundle the monodromies depend only on the cycle not on the starting point of the cycle.

THEOREM 5.3 ([**14**]). *For a $\mathbb{C}^2$-bundle on a finite graph with $SL_2(\mathbb{C})$ connection, we have*

$$\mathrm{qdet}\Delta = \sum_{CRSFsT} \prod_{e \in T} c_e \prod_{cycles\ of\ T} (2 - \mathrm{Tr}(w))$$

where the sum is over all CRSFs T, the first product is over the edges of T, and the second product is over the cycles of T, where w is the monodromy of the cycle (starting from some vertex, and in some arbitrary orientation).

Notice that the trace of the monodromy is independent of starting point (since conjugation does not change the trace) and orientation, since for a matrix $M \in \mathrm{SL}_2(\mathbb{C})$, we have $\mathrm{Tr}(M) = \mathrm{Tr}(M^{-1})$.

Here the function qdet is the **quaternion determinant** of the self-dual matrix Δ; this requires some explanation. A matrix M with entries in $GL_2(\mathbb{C})$ is said to be **self-dual** if $M_{ij} = \tilde{M}_{ji}$, where

$$\widetilde{\begin{pmatrix} a & b \\ c & d \end{pmatrix}} = \begin{pmatrix} d & -b \\ -c & a \end{pmatrix}.$$

Since $\Delta_{ij} = -c_{ij}\phi_{ij}$ where $\phi_{ij} \in \mathrm{SL}_2(\mathbb{C})$, we have $\Delta_{ji} = -c_{ij}\phi_{ij}^{-1} = \tilde{\Delta}_{ij}$ and so Δ is self-dual.

We define

$$\mathrm{qdet}(M) = \sum_{\sigma \in S_n} \mathrm{sgn}(\sigma) \prod_{\text{cycles}} \frac{1}{2}\mathrm{tr}(w)$$

where the sum is over the symmetric group, each permutation σ is written as a product of disjoint cycles, and $\mathrm{tr}(w)$ is the trace of the product of the matrix entries in that cycle. If we group together terms above with the same cycles—up to the order of traversal of each cycle—then the contribution from each of these terms is identical: reversing the orientation of a cycle does not change its trace. So we can write

$$\mathrm{qdet}(M) = \sum_{\text{cycle decomps}} (-1)^{c+n} \prod_{i=1}^{c} \widehat{\mathrm{tr}}(w_i),$$

where the sum is over cycle decompositions of the indices (not taking into account the orientation of the cycles), c is the number of cycles, and w_i is the monodromy (in one direction or the other) of each cycle. Here $\widehat{\mathrm{tr}}$ is equal to the trace for cycles of length at least 3; cycles of length 1 or 2 are their own reversals so we define $\widehat{\mathrm{tr}}(w) = \frac{1}{2}\mathrm{tr}(w)$ for these cycles.

As an example, let $A = aI, C = cI$ and $B = \begin{pmatrix} b_1 & b_2 \\ b_3 & b_4 \end{pmatrix}$. Then

$$\text{Qdet}\begin{pmatrix} A & B \\ \tilde{B} & C \end{pmatrix} = \widehat{\text{tr}}(A)\widehat{\text{tr}}(C) - \widehat{\text{tr}}(B\tilde{B}) = ac - (b_1b_4 - b_2b_3).$$

Note that if M is a self-dual $n \times n$ matrix then ZM, considered as a $2n \times 2n$ matrix is antisymmetric, where Z is the matrix with diagonal blocks $\begin{pmatrix} 0 & -1 \\ 1 & 0 \end{pmatrix}$ and zeros elsewhere. The following theorem allows us to compute Q-determinants explicitly.

THEOREM 5.4 ([**9**]). *Let M be an $n \times n$ self-dual matrix with entries in $GL_2(\mathbb{C})$ and M' the associated $2n \times 2n$ matrix, obtained by replacing each entry with the 2×2 block of its entries. Then* $\text{qdet}(M) = \text{Pf}(ZM')$, *the Pfaffian of the antisymmetric matrix ZM'.*

Note that the matrix Δ' (obtained by replacing entries in Δ with the 2×2 block of their coefficients) is just the matrix Δ acting on the total space of the bundle $W_{\mathcal{G}}$. So up to a sign we can write

$$\text{qdet}\Delta = (\det \Delta)^{1/2}.$$

5.3.1. *Unitary connections and measures.* In the case of a line bundle, if $\prod_{\text{cycles}} 2 - w - 1/w \geq 0$ for every CRSF, we can define, following Theorem 5.2, a probability measure on CRSFs in which a CRSF has probability proportional to the product of its edge weights times the factor $\prod_{\text{cycles}} 2 - w - 1/w$.

For example in the case that $|\phi_e| = 1$ for every edge, that is, ϕ is a **unitary connection**, then Δ is a Hermitian, positive semi-definite matrix (and positive definite for generic unitary connections). In this case $|w| = 1$ for all monodromies and so $2 - w - 1/w \geq 0$, and equal to zero only for cycles with trivial monodromy.

Similarly in the case of a two-dimensional bundle if $\phi_e \in \text{SU}_2$ for all edges then Δ is unitary and $2 - \text{Tr}w \geq 0$ for all monodromies.

In both cases we call $\mu = \mu_\Phi$ the corresponding probability measure.

Examples of natural settings of these measures are when the graphs are embedded on surfaces with geometric structures, such as a Riemannian metric (in which case the Levi-Civita connection defines parallel transport of vectors in the tangent bundle across an edge and provides a $U(1)$-connection) or an SU_2 structure on the surface and parallel transport restricted to the graph gives a SU_2 connection.

In the case of a flat connection, the monodromy of a contractible cycle is trivial, so that the probability measure is supported on CRSFs whose cycles are all topologically nontrivial. We call such CRSFs **essential CRSFs**.

5.4. Cycle-rooted groves. For the bundle Laplacian, the natural objects replacing groves are cycle-rooted groves. Given a graph $\mathcal{G}$ embedded on a surface with nodes $\mathcal{N}$, a **cycle-rooted grove** (CRG) is a collection of edges with the property that each component is either

(1) A tree containing one or more nodes
(2) a cycle-rooted tree containing no nodes.

Theorems 5.2 and 5.3 have an extension to the case of a graph with boundary $B \subset V$; the results are that the bundle Laplacian determinants (for the Laplacian with boundary) are weighted sums of cycle rooted groves; the weights of components with cycles are as before, and the weights of tree components are just the product of their edge weights (there is no "monodromy" contribution for tree components). See [**21**].

6. Annular networks

6.1. Laplacian determinant. Let $\mathcal{G}$ be a network with flat $\mathbb{C}^*$-connection on annulus, and with no nodes. Let z be the monodromy around a loop generating the homotopy group of the annulus. Then any simple closed loop on $\mathcal{G}$ will have monodromy 1 if contractible, z or z^{-1} if noncontractible.

As a consequence by Theorem 5.2 the Laplacian determinant is

$$\det \Delta = \sum_{k=1}^{\infty} C_k \left(2 - z - \frac{1}{z}\right)^k, \tag{16}$$

where C_k is the weighted sum of CRSFs having k components.

Thus $P(z) = \det \Delta$ is a Laurent polynomial in z from which one can extract the information about the number of loops in a random sample of a CRSF on $\mathcal{G}$. For example substituting $x = 2-z-1/z$ we have that $P(x)/P(1)$ is the probability generating function for the number of components.

Independently of the conductances, $P(z)$ contains topological information about the graph $\mathcal{G}$, for example the highest power of z is the maximum number of disjoint loops which can be drawn on $\mathcal{G}$, each winding around the annulus. This is because any such set of loops can be completed to a grove by adding some edges.

There is a simple characterization of the roots of P. By (16) the roots come in reciprocal pairs one of which is a double root at $z = 1$.

THEOREM 6.1. *P is reciprocal and the roots of P are real, positive, and distinct except for a double root at $z = 1$.*

PROOF. Our first goal is to minimize the graph so that it is a string of loops as illustrated in Figure 15, each loop winding around the annulus.

As in the proof of Theorem 5.1 above, we use the curve-shortening flow on the medial strands until we have only a packet of strands winding around the core of the annulus as in Figure 13, left (but without the vertical strands). In the resulting network, choose vertices v_1, v_2, one adjacent to each boundary and temporarily consider these to be nodes. There is one strand wrapping outside v_1 and one wrapping outside v_2. Break these strands so they have two stubs near each of v_1, v_2. Now continue to minimize the medial strands as in the proof of Theorem 5.1. In the minimal network, the stubs from v_1 must cross to the stubs of v_2 (rather than to each other), since otherwise the

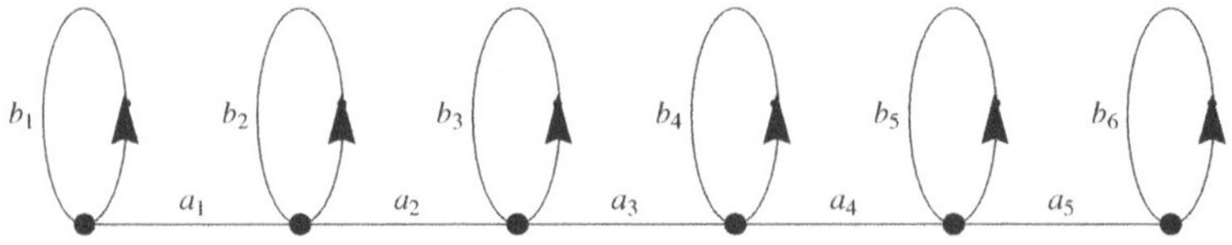

FIGURE 15. Every graph on an annulus can be minimized to a string of loops.

network would be disconnected. We can push all intersections of the finite strands (that is, those connecting stubs) to one side of the intersections of the stub/river strands, as described in Section 6.4.1 and Figure 20 below. The network and medial graph now resemble that in Figure 16. The graph can now easily be converted into a string of loops using the sequence of operations illustrated in Figure 17.

Let $a_1, \ldots, a_{n-1}$ be the conductances on the edges of the string and $b_1, \ldots, b_n$ the conductances on the loops. Then the Laplacian is a tridiagonal matrix (using $X = 2 - z - 1/z$)

$$\Delta = \begin{pmatrix} a_1 + b_1 X & -a_1 & 0 & & \\ -a_1 & a_1 + a_2 + b_2 X & -a_2 & 0 & \\ 0 & -a_2 & \ddots & & \\ & & & a_{n-2} + a_{n-1} + b_{n-1} X & -a_{n-1} \\ & & & -a_{n-1} & a_{n-1} + b_n X. \end{pmatrix}$$

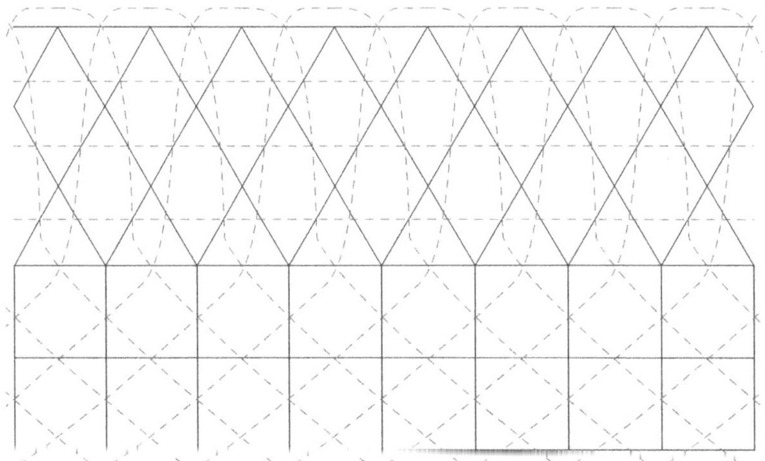

FIGURE 16. Intermediate stage in minimization of network on an annulus. (Medial strands dashed/red).

We pre- and post-multiply by the diagonal matrix U whose diagonal entries are $b_i^{-1/2}$ to get

$$U\Delta U = \begin{pmatrix} \frac{a_1}{b_1}+X & \frac{-a_1}{\sqrt{b_1b_2}} & 0 & & \\ \frac{-a_1}{\sqrt{b_1b_2}} & \frac{a_1}{b_2}+\frac{a_2}{b_2}+X & \frac{-a_2}{\sqrt{b_2b_3}} & 0 & \cdots \\ 0 & & \ddots & & \\ & & & \frac{a_{n-2}}{b_{n-1}}+\frac{a_{n-1}}{b_{n-1}}+X & \frac{-a_{n-1}}{\sqrt{b_{n-1}b_n}} \\ & & & \frac{-a_{n-1}}{\sqrt{b_{n-1}b_n}} & \frac{a_{n-1}}{b_n}+X. \end{pmatrix}$$

Thus we see that for each root z of P, $-2+z+\frac{1}{z}$ is an eigenvalue of

$$M = \begin{pmatrix} \frac{a_1}{b_1} & \frac{-a_1}{\sqrt{b_1b_2}} & 0 & & \\ \frac{-a_1}{\sqrt{b_1b_2}} & \frac{a_1}{b_2}+\frac{a_2}{b_2} & \frac{-a_2}{\sqrt{b_2b_3}} & 0 & \cdots \\ 0 & & \ddots & & \\ & & & \frac{a_{n-2}}{b_{n-1}}+\frac{a_{n-1}}{b_{n-1}} & \frac{-a_{n-1}}{\sqrt{b_{n-1}b_n}} \\ & & & \frac{-a_{n-1}}{\sqrt{b_{n-1}b_n}} & \frac{a_{n-1}}{b_n}. \end{pmatrix}.$$

This matrix M is symmetric and so has real eigenvalues, which are positive by (16). Symmetry also implies that M has orthogonal eigenvectors. Thus if $-X$ is a multiple eigenvalue of M it has an eigenspace V_{-X} of dimension ≥ 2. This means there is a nonzero eigenvector of M with eigenvalue $-X$ whose first coordinate is 0. But the coordinates of an eigenvector $(x_1, x_2, \ldots, x_n)$ satisfy a length-three linear recurrence:

$$\lambda x_1 = \frac{a_1}{b_1}x_1 - \frac{a_1}{\sqrt{b_1b_2}}x_2$$

and for $i > 1$

$$\lambda x_i = -\frac{a_{i-1}}{\sqrt{b_{i-1}b_i}}x_{i-1} + \left(\frac{a_{i-1}}{b_i}+\frac{a_i}{b_i}\right)x_i - \frac{a_i}{\sqrt{b_ib_{i+1}}}x_{i+1}.$$

Starting from $x_1 = 0$ this implies (using that all $a_i \neq 0$) that $x_i = 0$ for all i. This is a contradiction. We conclude that the eigenvalues of M, and therefore the roots of P, are distinct. □

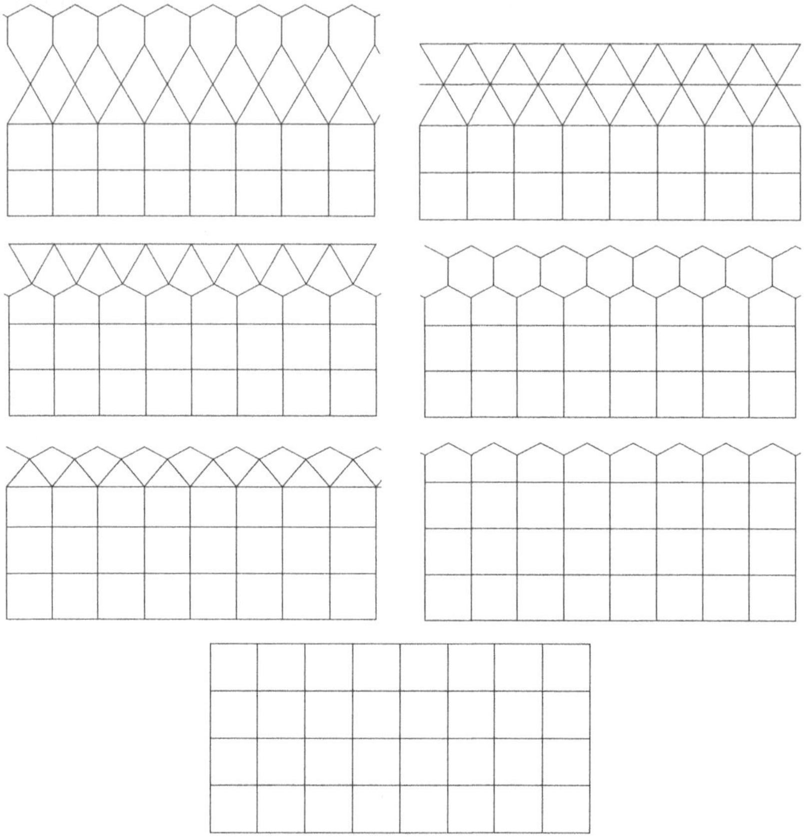

FIGURE 17. Passing from the intermediate stage to the string of loops: (1) From the network in Figure 16, use Delta-Y moves on the upper row of triangles. (1)-(2) Flatten the degree-2 vertices on the upper row and use Y-Deltas on the second row of vertices. (2)-(3) Use Delta-Ys on the row of triangles just above the squares. (3)-(4) Use Delta-Ys on the upper row of triangles. (4)-(5) Y-Delta. (5)-(6) Flatten and then Delta-Y. (6)-(7) Flatten.

6.2. Cylinder example. Let us compute, for a rectangular cylinder, $\det \Delta$ for the flat line bundle with monodromy z. This will allow us to compute the corresponding distribution of cycles in a uniform random essential CRSF. Let $H_{m,n}$ be the square grid network on a cylinder, obtained from the $n \times m$ square grid by adding edges from (n,i) to $(0,i)$ for each $i \in [1,m]$. $H_{m,n}$ is a "product" of an m-vertex linear network $\mathcal{G}_m$ and a circular network $\mathbb{Z}_n$ of length n. We put a flat line bundle structure on $H_{m,n}$ by putting parallel transport z on the edges $(n,i)(0,i)$ and 1 on all other edges.

The eigenvalues of the Laplacian $\Delta_{\mathcal{G}_m}$ on the linear network are $2 - 2\cos\frac{k\pi}{m}$ for $k = 1, 2, \ldots, m$ (the corresponding eigenvectors are $f_k(x) = \cos\frac{\pi k(x+1/2)}{m}$ for $x = 0, 1, 2, \ldots, m-1$).

The eigenvalues of the line-bundle Laplacian $\Delta_{\mathbb{Z}_n}$ on $\mathbb{Z}_n$ with monodromy z are $2 - \zeta - 1/\zeta$ where ζ is an nth root of z. (The corresponding eigenvectors are $g_k(x) = \zeta^x$.) The eigenvectors on $H_{m,n}$ are the products $f_k g_\ell$ for $(k, \ell) \in [1, m] \times [0, n-1]$.

We then have

$$\det \Delta_\Phi = \prod_{\zeta^n = z} \prod_{k=1}^{m} 4 - 2\cos\frac{k\pi}{m} - \zeta - \frac{1}{\zeta}.$$

Use the identity

$$\prod_{\zeta^n = z} R - \zeta - \frac{1}{\zeta} = \mathrm{Ch}_n(R)$$

where Ch_n is defined by $\mathrm{Ch}_n(\alpha + \frac{1}{\alpha}) = \alpha^n + \alpha^{-n}$ (a variant of the Chebychev polynomial).

This leads to

$$\begin{aligned}\det \Delta &= (2 - z - z^{-1}) \prod_{k=1}^{m} \left(\mathrm{Ch}_n\left(4 - 2\cos\frac{k\pi}{m}\right) - z - \frac{1}{z}\right)\\ &= w \prod_{k=1}^{m} w + \mathrm{Ch}_n\left(4 - 2\cos\frac{k\pi}{m}\right) - 2\end{aligned}$$

where $w = 2 - z - \frac{1}{z}$. If we let $Q(w)$ be this polynomial then $Q(w)/Q(1)$ is the probability generating function of the number of cycles in a uniform random essential CRSF.

It is interesting to consider what happens to this distribution for a large annulus, when $m, n \to \infty$ with m/n converging to a fixed quantity τ. For large n, $\mathrm{Ch}_n(4 - 2\cos\frac{k\pi}{m})$ is large unless k is near 0. Thus only values of k near 0 affect the limiting distribution. We have $4 - 2\cos\frac{k\pi}{m} = 2 + \frac{\pi^2 k^2}{m^2} + O(\frac{k}{m})^4 = \alpha_k + 1/\alpha_k$ where $\alpha_k = 1 + \frac{\pi k}{m} + O(\frac{k}{m})^2$. Thus in the limit $m, n \to \infty$ with $m/n \to \tau$ we have $\mathrm{Ch}_n(4 - 2\cos\frac{k\pi}{m}) = 2\cosh\frac{\pi k}{\tau} + o(1)$. The limit probability generating function for the number of cycles is then

$$Q(w)/Q(1) = w \prod_{k=1}^{\infty} \left(\frac{w + 2\cosh\frac{\pi k}{\tau} - 2}{2\cosh\frac{\pi k}{\tau} - 1}\right).$$

See Figure 18.

6.3. The response matrix. Lam and Pylyavskyy [**24**] studied the response matrices of networks on annuli. Their point of view was to consider the lift $\tilde{\mathcal{G}}$ of the network $\mathcal{G}$ to the universal cover of the annulus. There one can define the response matrix as the limit of response matrices on larger and larger portions of the graph; it is not hard to show that this

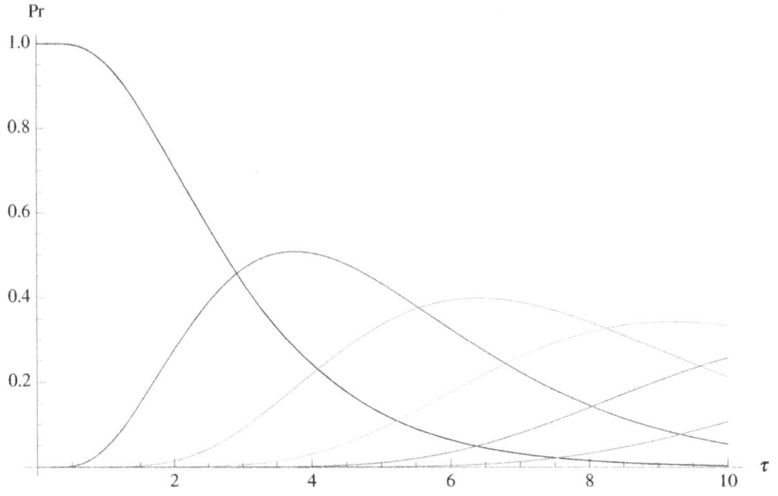

FIGURE 18. The probabilities of $1, 2, \ldots, 6$ cycles in a uniform random essential CRSF on a large annulus, as a function of the modulus τ.

limit exists. Since the universal cover is planar one can recover some of the results from the planar case like nonnegativity of the non-interlaced minors (non-interlacedness makes sense since the lift $\tilde{\mathcal{G}}$ is planar).

We will take a different approach here which uses the bundle Laplacian for a flat connection. Let $\mathcal{G}$ be a network drawn on an annulus Σ, with nodes $\mathcal{N}$ which are a subset of the vertices adjacent to the two boundaries of Σ. Let Φ be a flat $\mathbb{C}^*$-connection with monodromy z around a generator of $\pi_1(\Sigma)$. Let Δ be the associated bundle Laplacian. Let L be the response matrix, defined as before as

$$L = -A + BC^{-1}B^*$$

where $\Delta = \begin{pmatrix} A & B \\ B^* & C \end{pmatrix}$. Now entries in L are rational functions of z (with coefficients which depend on the conductances).

What can be said about EIT on annular networks?

For circular planar networks it was very useful to have combinatorial interpretations (in terms of groves) of the entries and minors of the response matrix. We have similar interpretations in the present case. The following theorem of **[21]** holds for any network with line bundle (not just networks on an annulus). Compare Theorem 4.3 above.

THEOREM 6.2 ([**21**]) *Let Q, R, S, T be a partition of $\mathcal{N}$ and $|R| = |S|$. Then $\det L_{RUT}^{SUT}$ is the ratio of two terms: the denominator is the weighted sum of CRSFs; the numerator is a signed weighted sum of cycle-rooted groves of $\mathcal{G}_T$, the graph $\mathcal{G}$ in which all nodes in T are considered internal, the nodes in Q are in singleton parts, and in which nodes in R are paired with nodes in*

S, with the sign being the sign of the pairing permutation, times the parallel transports from S to R:

$$\det L_{R,T}^{S,T} = \sum_{\textit{permutations } \rho} (-1)^{\rho} \frac{\mathcal{Z}\left[\substack{s_{\rho(1)} \\ r_1}\middle|\cdots\middle|\substack{s_{\rho(k)} \\ r_k}\middle|q_1\middle|\cdots\middle|q_\ell\right]}{\mathcal{Z}[1|2|\cdots|n]}$$

Here the numerator $\mathcal{Z}\left[\substack{s_{\rho(1)} \\ r_1}\middle|\cdots\middle|\substack{s_{\rho(k)} \\ r_k}\middle|q_1\middle|\cdots\middle|q_\ell\right]$ is the weighted sum of CRG's in which there are tree components connecting r_i to $s_{\rho(i)}$ for each i; the weight is the product of the edge conductances, times the product over all cycles γ of $2 - w(\gamma) - 1/w(\gamma)$ (where $w(\gamma)$ is the monodromy of the cycle γ) and times the product of the parallel transports from the r_i to the $s_{\rho(i)}$ along the edges of the trees. We use the notation $\mathcal{Z}$ as opposed to Z to remind us that we are dealing with the bundle Laplacian.

On an annulus with flat connection, it is convenient to have the connection supported on a "zipper", that is, the set of faces crossing a shortest path in the dual graph from one boundary component to the other, as in Figure 19. In such a case the monodromy along a path from a node to another node is $1, z$ or z^{-1} if the nodes are on the same boundary; it is a power of z if the nodes are on different boundaries, and it is $z^{\pm 1}$ for a topologically nontrivial cycle.

If particular if i and j are nodes on opposite boundaries then

$$L_{ij}\mathcal{Z}_{\text{unc}} = \sum_{k\in\mathbb{Z}} C_k z^k,$$

where C_k is the weighted sum of CRGs with a component tree connecting i to j, no cycles (a cycle is precluded by the existence of a path from i to j), and where k is the signed number of times the path from i to j crosses the zipper.

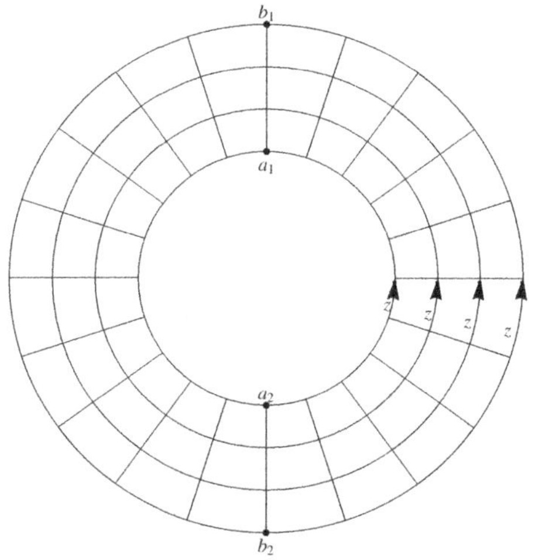

FIGURE 19. The parallel transport on an annulus can be supported on a zipper.

Thus $L_{ij}\mathcal{Z}_{\text{unc}}$ is a Laurent polynomial of z with nonnegative coefficients.

Here is another example. Suppose $A = \{a_1, a_2\}$ are nodes on one boundary and $B = \{b_1, b_2\}$ are nodes on the other. Then by the Theorem $L_A^B\mathcal{Z}_{\text{unc}}$ contains terms in which a_1 is paired with b_1 (and a_2 with b_2) and, with an opposite sign terms in which a_1 is paired with b_2 and a_2 with b_1. However these two types of terms differ by an power of z. Suppose for example that the zipper is as shown in Figure 19. Then any term ${}^{b_1}_{a_1}|^{b_2}_{a_2}$ has an even power of z and ${}^{b_2}_{a_1}|^{b_1}_{a_2}$ has an odd power of z. Thus $L_A^B\mathcal{Z}_{\text{unc}}$ is a Laurent polynomial in z with coefficients of alternating sign.

6.4. Minimality and reconstruction.

6.4.1. *Minimal networks on the annulus: landscapes.* Let $\mathcal{G}$ be a minimal network on an annulus Σ with n_1 nodes on one boundary and n_2 on the other. As discussed, let $\tilde{\mathcal{G}}$ be the lift of $\mathcal{G}$ to the strip, the universal cover of the annulus. The medial strands come in four types:

(1) they connect stubs on the lower boundary of the strip (the **rocks**);
(2) they connect stubs on the upper boundary of the strip (the **clouds**);
(3) they connect stubs across the boundaries (the **trees**);
(4) they form bi-infinite paths along the strip (the **river**).

We can perform Y-Delta transformations so that the strands of the first, second and fourth type do not intersect (that is, no strand of the first type intersects a strand of the second or fourth type, although it may intersect other strands of the first type, etc.) As long as there is at least one tree strand, then we can "untwist" the river strands so that they are all parallel and noncrossing.

The picture is then like a landscape; the rocks at the bottom, interspersed with trees which grow up to mingle with the clouds, and the river runs through the trees but does not touch the clouds or rocks. See Figures 20 and 21.

6.4.2. *Reconstruction.* The reconstruction of the conductances on a minimal annular network as a function of the response matrix is open in general. Lam and Pylyavskyy [**24**] worked out the reconstruction for the "grid" network of Figure 22, whose medial graph is a grid.

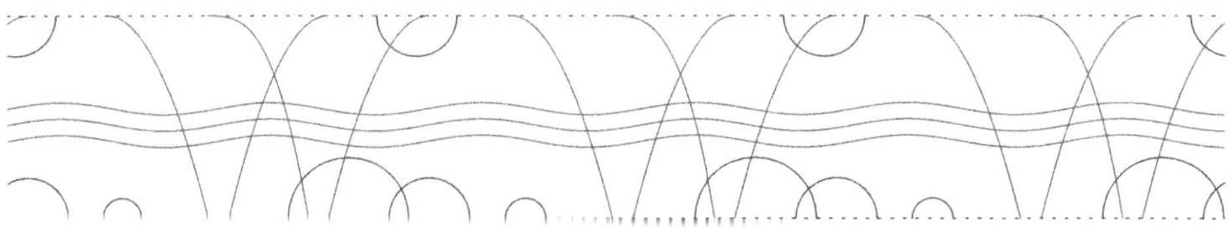

FIGURE 20. A landscape: the general form of a minimal network on an annulus (we show only the medial graph, and on the universal cover of the annulus).

FIGURE 21. Bow falls, Banff.

The new feature of this reconstruction problem is that the solution is not unique. If the medial graph of $\mathcal{G}$ has k topologically nontrivial cycles, then there are generically $k!$ solutions to the reconstruction. The idea is that given a set of conductances on $\mathcal{G}$ one can permute the parallel strands of the medial graph and get another solution. How does one permute two strands? This follows from the discussion of combing in Theorem 5.1 above. See Figure 13. One introduces a pair of edges of opposite conductance $c, -c$; this has no effect on the current flow or response matrix. Now perform Y-Delta moves around the annulus to exchange the adjacent strands. There is a unique choice of c for which the added edge, when it comes back around, will again have conductance c; it then can be removed along with the parallel edge of conductance $-c$.

In [**24**] it is conjectured that reconstruction is possible on all minimal annular networks.

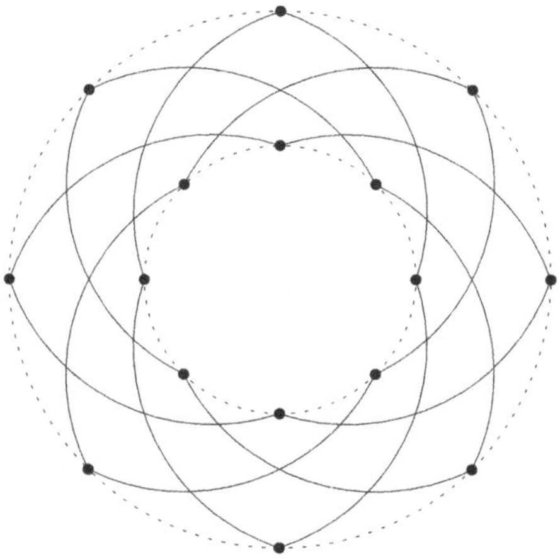

FIGURE 22. The "grid" network on an annulus.

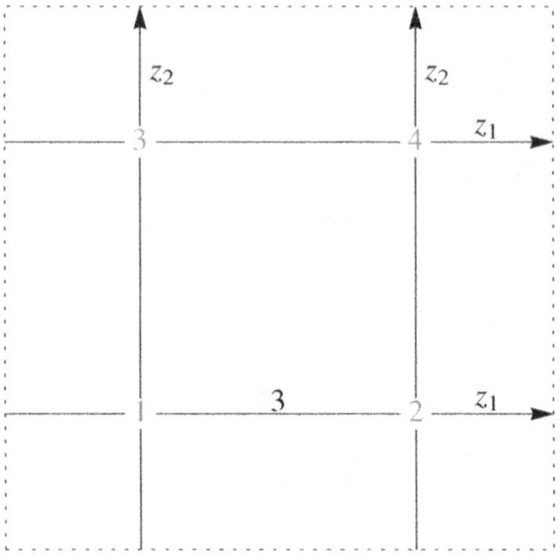

FIGURE 23. The 2×2 grid on a torus; in this example all edges have conductance 1 except for the one indicated which has conductance 3.

7. Periodic networks and networks on the torus

Let $\mathcal{G}$ be a network on a torus. By this we mean a network $\mathcal{G}$, with no boundary, embedded on a torus in such a way that every complementary component is contractible.

For concreteness we suppose the torus is $\mathbb{T}^2 = \mathbb{R}^2/\mathbb{Z}^2$ (but the network can be arbitrary). The lifted network $\tilde{\mathcal{G}}$ on the universal cover $\mathbb{R}^2$ of $\mathbb{T}^2$ is a periodic planar network.

This case differs from the previous cases of circular planar and annular networks in that the underlying surface has no boundary, so we do not introduce any nodes. Remarkably we can still have a complete theory about reducibility, minimal networks, and the reconstruction problem.

7.1. The spectral curve of Δ. Let Φ be a flat line bundle on $\mathcal{G}$ with monodromy z_1, z_2 around the standard generators of the homotopy group of $\mathbb{T}^2$. Let $\Delta = \Delta_\Phi$ the associated Laplacian. See for example Figure 23. In this example the associated Laplacian determinant $P(z_1, z_2) = \det \Delta$ is (with the vertices in the order indicated)

$$P(z_1,z_2) = \det \begin{pmatrix} 6 & -3-z_1^{-1} & -1-z_1^{-1} & 0 \\ -3-z_1 & 6 & 0 & -1-z_2^{-1} \\ -1-z_2 & 0 & 4 & -1-z_1^{-1} \\ 0 & -1-z_2 & -1-z_1 & 4 \end{pmatrix}$$

$$= 3z_1^2 + \frac{3}{z_1^2} - 4z_1z_2 - \frac{4z_1}{z_2} - \frac{4z_2}{z_1} - \frac{4}{z_1z_2} - 76z_1 - \frac{76}{z_1} + z_2^2$$

$$+ \frac{1}{z_2^2} - 52z_2 - \frac{52}{z_2} + 264. \tag{17}$$

In general P is a (Laurent) polynomial with coefficients which are polynomials in the conductances. It is called the **characteristic polynomial** of Δ. The curve $\{(z_1, z_2) \in \mathbb{C}^2 \mid P(z_1, z_2) = 0\}$ is the **spectral curve** of Δ. Note that P is symmetric: $P(z_1, z_2) = P(z_1^{-1}, z_2^{-1})$. This is because we can write $\Delta = d^*\mathcal{C}d$ where d^* is the Hermitian transpose of d when $|z_1| = |z_2| = 1$.

The **Newton polygon** of the network $\mathcal{G}$ on the torus is the Newton polygon of P, that is, the convex hull of

$$\{(i, j) \in \mathbb{Z}^2 \mid z_1^{i_1} z_2^{j_2} \text{ is a coefficient of } P\}.$$

It is a centrally symmetric polygon centered at $(0, 0)$.

By Theorem 5.2 we can write

$$\begin{aligned} P(z_1, z_2) &= \sum_{\text{CRSFs}} \left(\prod_e c_e\right) (2 - w - 1/w)^k, \\ &= \sum_{(r,s)\neq(0,0)} C_{r,s}(2 - z_1^i z_2^j - z_1^{-i} z_2^{-j})^k, \end{aligned} \tag{18}$$

where in the first line we used the fact that in a CRSF on a torus all cycles have the same homology class, and in the second line we sum over all homology classes $(r, s) \in \mathbb{Z}^2$ (we need only sum over one of each pair $(r, s), (-r, -s)$) where $(r, s) = (ik, jk)$ with i, j being relatively prime. The coefficient $C_{r,s}$ is the weighted sum of CRSFs with k cycles of homology class (i, j), that is (if we orient the cycles in the same direction) of total homology class (r, s).

This shows that the $z_1^r z_2^s$ coefficient of $P(z_1, z_2)$ is only due to the CRSFs with cycles in direction r/s. Thus the Newton polygon N tells us which homology classes of CRSFs are possible for $\mathcal{G}$; equivalently, the maximal number of vertex-disjoint cycles in $\mathcal{G}$ which have homology class (i, j) is determined by the integer point in N which is farthest from the origin in direction (i, j).

In the above example (17), P can be written in "Newton polygon form"

$$P = \begin{matrix} & & 1 & & \\ & -4 & -52 & -4 & \\ 3 & -76 & 264 & -76 & 3. \\ & -4 & -52 & -4 & \\ & & 1 & & \end{matrix}$$

From this, one can write P in a unique way in form (18):

$$\begin{aligned} P(z_1, z_2) = 3\left(2 - z_1 - \frac{1}{z_1}\right)^2 + 64\left(2 - z_1 - \frac{1}{z_1}\right) + \left(2 - z_2 - \frac{1}{z_2}\right)^2 \\ + 48\left(2 - z_2 - \frac{1}{z_2}\right) + 4\left(2 - z_1 z_2 - \frac{1}{z_1 z_2}\right) + 4\left(2 - \frac{z_1}{z_2} - \frac{z_2}{z_1}\right). \end{aligned} \tag{19}$$

Thus there are, for example, exactly 48 CRSFs with a single cycle which has homology class $(0, \pm 1)$.

7.2. UST on the plane. Let us say briefly a few words about the UST on the plane.

Let $\tilde{\mathcal{G}}$ be the periodic planar graph which is the lift of $\mathcal{G}$ to $\mathbb{R}^2$. Pemantle **[28]** showed that there is a UST measure on spanning trees of $\tilde{\mathcal{G}}$ which is the limit of the UST measure on larger and larger tori $\tilde{\mathcal{G}}_n \stackrel{\text{def}}{=} \tilde{\mathcal{G}}/n\mathbb{Z}^2$.

The **Free Energy** of this measure is the limit

$$F = \lim_{n\to\infty} \frac{1}{n^2} \log Z_n,$$

where Z_n is the partition sum for spanning trees on $\tilde{\mathcal{G}}_n$.

There is a nice formula for F

$$F = \frac{1}{(2\pi i)^2} \int_{|z_1|=1} \int_{|z_2|=1} \log P(z_1, z_2) \frac{dz_1}{z_1} \frac{dz_2}{z_2}$$

which arises from the exact formula

$$Z_n(z_1, z_2) = \prod_{\zeta_1^n = z_1} \prod_{\zeta_2^n = z_2} P(\zeta_1, \zeta_2)$$

(obtained by Fourier analysis of $\tilde{\mathcal{G}}_n$) by taking logs and replacing the Riemann sum with an integral.

One can analyze the limiting UST measure using the fact that edge process is determinantal with kernel T given by the **Transfer Current** matrix: $T(e, e')$ is the current across e' when one unit of current enters the graph at e_+ and exits at e_-. In terms of the Greens function we can write

$$T(e, e') = G(e_+, e'_+) - G(e_+, e'_-) - G(e_-, e'_+) + G(e_-, e'_-).$$

The Green's function $G(v, v')$ on $\tilde{\mathcal{G}}$ has a formula involving the Fourier coefficients of $1/P(z_1, z_2)$: more precisely, an integral over the unit torus $|z_1| = |z_2| = 1$ of an expression $z_1^x z_2^y Q(z_1, z_2)/P(z_1, z_2)$ where (x, y) is the translation from v to v' and Q is a polynomial depending on where v, v' sit in their respective fundamental domains. In particular the transfer current has an exact asymptotic expression.

7.3. Medial strands and minimality. Recall that two networks on $\mathbb{T}^2$ are **topologically equivalent** if one can be obtained from the other (disregarding conductances) by electrical transformations. We say that they are **electrically equivalent** if their characteristic polynomials are the same (compare this definition with the one for networks with boundary in Section 4.4.2).

THEOREM 7.1. *Two networks on a torus are topologically equivalent if and only if their Newton polygons are equal.*

PROOF. It is not hard to show that electrical transformations do not change N. The simplest reason is because there is a bijection on the CRSFs of the "before" and "after" networks, and the set of CRSFs determines N.

So it suffices to show that N determines a minimal network up to electrical equivalence. Suppose N has $2n$ integer boundary points $v_0, \ldots, v_{2n-1}$ in cclw order. Let e_i be the edge joining v_i to v_{i+1}; it is antiparallel with e_{n+i} because P is centrally symmetric. The proof now follows from the following lemma. □

LEMMA 7.2. *The strands of a minimal network are in bijection with, and have homology classes equal to, the n edge pairs $\{e_i, e_{i+n}\}$.*

PROOF. Since the network is minimal we can isotope the strands on the torus so that they are geodesics for a Euclidean structure. Then we orient the strands so that their y coordinate is increasing, or if they have no y component, the x component is increasing.

Let $S_1, \ldots, S_k$ be the strands, and let $e_i = (x_i, y_i) \in \mathbb{Z}^2$ be the homology class of S_i. The length $2k_i$ of S_i (that is, the number of edges of the corresponding zig-zag path of $\mathcal{G}$) is then determined by $2k_i = \sum_j |e_i \wedge e_j|$ since exactly two strands cross at each edge of $\mathcal{G}$.

Every strand gives a constraint on the homology class of a CRSF: for any CRSF with homology class ω we must have

$$|\omega \wedge e_i| \leq k_i, \tag{20}$$

since $|\omega \wedge e_i|$ is the intersection number of the CRSF with the strand S_i.

Suppse the S_i are indexed in order of decreasing slope x_i/y_i. Let N_S be the convex polygon with side vectors $e_1, \ldots, e_k, -e_1, \ldots, -e_k$ in cclw order. It has vertices $v_1, v_2, \ldots, v_k, -v_1, -v_2, \ldots, -v_k$ where

$$v_i = \frac{1}{2}(e_1 + \cdots + e_{i-1} - e_i - \cdots - e_k).$$

We claim that N_S is the subset of $\omega \in \mathbb{R}^2$ which satisfy the constraints (20). To see this, note that if v_i, v_{i+1} are consecutive vertices of N_S and e_i is the vector between them, then

$$\begin{aligned} 2v_i \wedge e_i &= e_1 \wedge e_i + \cdots + e_{i-1} \wedge e_i - e_{i+1} \wedge e_i - \cdots - e_k \wedge e_i \\ &= \sum_j |e_j \wedge e_i| \\ &\leq 2k_i \end{aligned}$$

and similarly for v_{i+1}. Thus v_i, v_{i+1} satisfy the maximal possible constraint.

Thus $N \subset N_S$. To see that $N = N_S$, is suffices to construct a CRSF with homology class equal to each of the vertices v_i of N_S. To construct a set of cycles with homology class $v_1 = \frac{1}{2}(-e_1 - e_2 - \cdots - e_k)$, for example, put weight $-1/2$ on each edge of the zig-zag path of S_i (in the orientation defined above). Each edge of $\mathcal{G}$ will then have weight 0 or ± 1; since we assumed that the strands were drawn geodesically, at each vertex there is

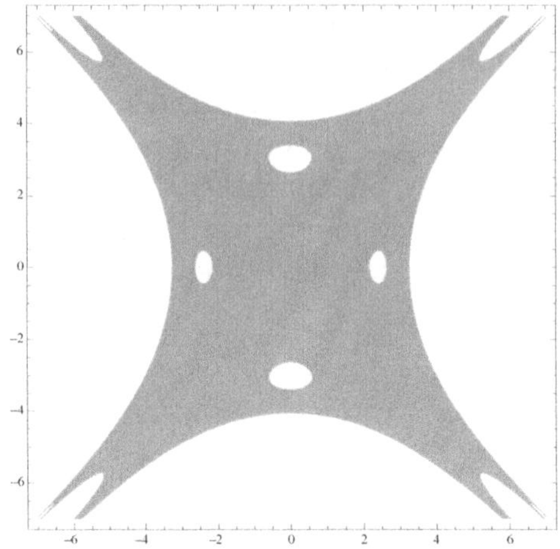

FIGURE 24. The amoeba (shaded) of the curve of Figure 23 above. The boundary of the amoeba is the real locus of $P = 0$; a model of $P = 0$ can be obtained by gluing two copies of the shaded region together along their common boundary.

exactly one incoming and one outgoing edge. The oriented edges with weight 1 form a set of vertex disjoint cycles of the homology class ω. □

7.4. Harnack curves and characterization of spectral curves. The Laurent polynomial $P = \det \Delta$ has the following properties:

(1) P has real coefficients.
(2) P is symmetric: $P(z_1, z_2) = P(z_1^{-1}, z_2^{-1})$.
(3) $P(1,1) = 0$.
(4) $\{P = 0\}$ is a Harnack curve.

What we call Harnack curves were studied by Harnack and called simple Harnack curves by Mikhalkin in [**26**]. The definition given by Mikhalkin in [**26**] involves a topological condition on the real locus; in [**27**] a more concise characterization was given: a curve is (simple) Harnack if it intersects each torus $\{|z| = c_1, |w| = c_2\}$ in at most two points (which are necessarily complex conjugates if not real). In particular the map $(z, w) \mapsto (\log|z|, \log|w|)$ which maps the curve into $\mathbb{R}^2$ is generically 2 to 1. The image of $P = 0$ under this map is the **amoeba** of P. See Figure 24 for the amoeba of the example above.

Remarkably, the four properties above characterize the characteristic polynomials of networks on a torus:

THEOREM 7.3. [**15, 12**] *For any polynomial P satisfying the above properties there is a network $\mathcal{G}$ whose characteristic polynomial is P. The set of minimal networks with polynomial P, modulo electrical transformations, is a real torus of dimension $g/2$, where g is the geometric genus (the number of real ovals of the amoeba of P).*

8. Other surfaces

For surfaces other than the annulus and torus, the fundamental group will be nonabelian so it is appropriate to use an $\mathrm{SL}_2(\mathbb{C})$-connection on $\mathcal{G}$ rather than a $\mathbb{C}^*$-connection.

Although we understand very little about this situation, there is one positive result. Considering the Laplacian determinant as a function on the moduli space of flat $\mathrm{SL}_2(\mathbb{C})$-connections, it can be written in a unique way as a sum over CRSFs according to the homotopy types of their cycles. Let us explain this. We define a **finite lamination** on Σ to be an isotopy class of finite pairwise disjoint collection of simple closed curves, none bounding a disk. For example on an annulus there is a finite lamination for each nonnegative integer k, k being the number of its cycles.

The cycles in a CRSF form a finite lamination. One can ask the question: for a given finite lamination $\mathcal{L}$, what is the weighted sum of CRSFs having cycle set of type $\mathcal{L}$? We have

THEOREM 8.1 ([**14, 10**]). *$P = \det\Delta$ can be written as*

$$P = \sum_{\mathcal{L}} C_{\mathcal{L}}$$

where the sum is over all finite laminations and $C_{\mathcal{L}}$ is the weighted sum (partition sum) of CRSFs of cycle type $\mathcal{L}$. The coefficients $C_{\mathcal{L}}$ are functions of P only, and can be extracted via an appropriate integration of P over the representation variety $Hom(\pi_1(\Sigma), SL_2(\mathbb{C}))$.

As an example, take Σ to be a pair of pants, that is, a sphere minus three disks. Let $\mathcal{G}$ be a network embedded on Σ. Take a flat $\mathrm{SL}_2(\mathbb{C})$-connection on $\mathcal{G}$ with monodromy A, B, C around the three boundary holes, where $ABC = 1$.

Then

$$P = \det\Delta = \sum_{i,j,k\geq 0} c_{i,j,k}(2-\mathrm{Tr}A)^i(2-\mathrm{Tr}B)^j(2-\mathrm{Tr}(AB))^k.$$

Given variables X, Y, Z one can choose matrices A, B in $\mathrm{SL}_2(\mathbb{C})$ such that $2-\mathrm{Tr}A = X, 2-\mathrm{Tr}B = Y, 2-\mathrm{Tr}AB = Z$. Then

$$P = P(X,Y,Z) = \sum_{i,j,k\geq 0} c_{i,j,k}X^iY^jZ^k.$$

One can extract from this expression the coefficients $c_{i,j,k}$.

References

[1] Aldous, D. The random walk construction of uniform spanning trees and uniform labelled trees. SIAM J. Disc. Math. 3 (1990) 450–465.

[2] Broder, A. (1988). Generating random spanning trees. In: Symp. foundations of computer sci., Institute for Electrical and Electronic Engineers, New York 442–447.

[3] Borchardt, Carl Wilhelm Über eine Interpolationsformel für eine Art Symmetrischer Functionen und über Deren Anwendung. Math. Abh. der Akademie der Wissenschaften zu Berlin (1860) 1–20.
[4] Burton, Robert; Pemantle, Robin Local characteristics, entropy and limit theorems for spanning trees and domino tilings via transfer-impedances. Ann. Probab. 21 (1993), no. 3, 1329–1371.
[5] Cayley, Arthur A theorem on trees. Quart. J. Math 23 (1889), 376–378.
[6] Colin de Verdire, Yves Réseaux électriques planaires. I. Comment. Math. Helv. 69 (1994), no. 3, 351–374
[7] Colin de Verdire, Yves; Gitler, Isidoro; Vertigan, Dirk, Réseaux électriques planaires. II. (Comment. Math. Helv. 71 (1996), no. 1, 144–167
[8] Curtis, E. B.; Ingerman, D.; Morrow, J. A. Circular planar graphs and resistor networks. Linear Algebra Appl. 283 (1998), no. 1–3, 115–150
[9] Dyson, Freeman J. Correlations between eigenvalues of a random matrix. Comm. Math. Phys. 19 1970 235–250.
[10] Fock, Vladimir; Goncharov, Alexander Moduli spaces of local systems and higher Teichmller theory. Publ. Math. Inst. Hautes tudes Sci. No. 103 (2006), 1–211.
[11] Forman, Robin Determinants of Laplacians on graphs. Topology 32 (1993), no. 1, 35–46
[12] Goncharov, Alexander B.; Kenyon, Richard, Dimers and cluster integrable systems. arxiv:1107.55–88.
[13] Kenyon, Richard Dominos and the Gaussian free field. Ann. Probab. 29 (2001), no. 3, 1128–1137.
[14] Kenyon, Richard Spanning forests and the vector bundle laplacian, arXiv:1001.4028
[15] Kenyon, Richard; Okounkov, Andrei Planar dimers and Harnack curves. Duke Math. J. 131 (2006), no. 3, 499–524
[16] Kenyon, Richard; Okounkov, Andrei; Sheffield, Scott Dimers and amoebae. Ann. of Math. (2) 163 (2006), no. 3, 1019–1056
[17] Kenyon, Richard W.; Propp, James G.; Wilson, David B. Trees and matchings. Electron. J. Combin. 7 (2000), Research Paper 25, 34 pp. (electronic)
[18] Kenyon, Richard W.; Sheffield, Scott Dimers, tilings and trees. J. Combin. Theory Ser. B 92 (2004), no. 2, 295–317.
[19] Kenyon, Richard W.; Wilson, David B. Boundary partitions in trees and dimers. Trans. Amer. Math. Soc. 363 (2011), no. 3, 1325–1364.
[20] Kenyon, Richard W.; Wilson, David B. Combinatorics of tripartite boundary connections for trees and dimers. Electron. J. Combin. 16 (2009), no. 1
[21] Kenyon, Richard W.; Wilson, David B., Positivity in planar networks, in preparation.
[22] Kirchhoff, G. Über die Auflösung der Gleichungen, auf welche man bei der untersuchung der linearen verteilung galvanischer Ströme geführt wird. Ann. Phys. Chem. 72, 497–508, 1847.
[23] Lam, Thomas; Pylyavskyy, Pavlo, Electrical networks and Lie theory. arXiv:1103.3475.
[24] Lam, Thomas; Pylyavskyy, Pavlo Inverse problem in cylindrical electrical networks arXiv:1104.4998.
[25] Lawler, Gregory F.; Schramm, Oded; Werner, Wendelin Conformal invariance of planar loop-erased random walks and uniform spanning trees. Ann. Probab. 32 (2004), no. 1B, 939–995.
[26] Mikhalkin, Grigory, Real algebraic curves, the moment map and amoebas", Ann. Math. 151 (2000), 309–326.
[27] Mikhalkin, Grigory; Rullgärd, Hans Amoebas of maximal area. Internat. Math. Res. Notices 2001, no. 9, 441–451.
[28] Pemantle, Robin Choosing a spanning tree for the integer lattice uniformly. Ann. Probab. 19 (1991), no. 4, 1559–1574.

[29] Postnikov, Alexander, Total positivity, Grassmannians, and networks, arXiv:math/0609764.

[30] Talaska, Kelli Combinatorial formulas for Γ-coordinates in a totally nonnegative Grassmannian. J. Combin. Theory Ser. A 118 (2011), no. 1, 58–66.

[31] H. N. V. Temperley. In Combinatorics: Proceedings of the British Combinatorial Conference 1973, pages 202–204, 1974. London Mathematical Society Lecture Notes Series 13.

[32] Wilson, David Bruce Generating random spanning trees more quickly than the cover time. Proceedings of the Twenty-eighth Annual ACM Symposium on the Theory of Computing (Philadelphia, PA, 1996), 296–303, ACM, New York, 1996.

Current Developments in Mathematics, 2011

Hyperbolicity and stable polynomials in combinatorics and probability

Robin Pemantle[1,2]

ABSTRACT. These lectures survey the theory of hyperbolic and stable polynomials, from their origins in the theory of linear PDE's to their present uses in combinatorics and probability theory.

CONTENTS

1991 *Mathematics Subject Classification.* Primary: 26C10, 62H20; secondary: 30C15, 05A15.

Key words and phrases. amoeba, cone, dual cone, Gårding-hyperbolicity, generating function, half-plane property, homogeneous polynomial, Laguerre–Pólya class, multiplier sequence, multi-affine, multivariate stability, negative dependence, negative association, Newton's inequalities, Rayleigh property, real roots, semi-continuity, stochastic covering, stochastic domination, total positivity.

[1]Supported in part by National Science Foundation grant # DMS 0905937.

[2]University of Pennsylvania, Department of Mathematics, 209 S. 33rd Street, Philadelphia, PA 19104 USA, pemantle@math.upenn.edu.

1. Introduction

These lectures concern the development and uses of two properties, **hyperbolicity** and **stability**. Hyperbolicity has been used chiefly in geometric and analytic contexts, concerning wave-like partial differential equations, lacunas, and generalized Fourier transforms. Stability has its origins in control theory but has surfaced more recently in combinatorics and probability, where its algebraic properties, such as closure under various operations, are paramount. We begin with the definitions.

DEFINITION 1.1 (hyperbolicity). *A homogeneous complex d-variable polynomial $p(z_1, \ldots, z_d)$ is said to be hyperbolic in direction $\mathbf{x} \in \mathbb{R}^d$ if and only if*

$$p(\mathbf{x} + i\mathbf{y}) \neq 0 \text{ for all } \mathbf{y} \in \mathbb{R}^d . \tag{1.1}$$

A non-homogeneous d-variable polynomial q with leading homogeneous part p is said to be hyperbolic if and only if $p(\mathbf{x}) \neq 0$ and there is some $t_0 > 0$ such that

$$q(it\mathbf{x} + \mathbf{y}) \neq 0 \text{ for all } \mathbf{y} \in \mathbb{R}^d \text{ and real } t < t_0 . \tag{1.2}$$

A polynomial is said to be hyperbolic if it is hyperbolic in direction ξ for some $\xi \in \mathbb{R}^d$.

DEFINITION 1.2 (stability). *The complex polynomial q is said to be **stable** if and only if*

$$\Im\{z_j\} > 0 \text{ for all } j = 1, \ldots, d \text{ implies } q(z_1, \ldots, z_d) \neq 0 . \tag{1.3}$$

A ***real stable*** *polynomial is one that is both real and stable.*

The following relation holds between these two definitions.

PROPOSITION 1.3. *A real homogeneous polynomial p is stable if and only if it is hyperbolic in direction ξ for every ξ in the positive orthant.*

Hyperbolicity is a real geometric property, that is, it is invariant under the general linear group. It is not invariant under complex linear transformations, rather it distinguishes the real subspace of $\mathbb{C}^d$. Stability is invariant under yet fewer maps than is hyperbolicity, due to the fact that it distinguishes positive from negative. Stability is invariant under coordinatewise transformations of the upper half-plane, such as inversion $z \mapsto -1/z$ and dilation $z \mapsto \lambda z$, from which one can get a surprising amount of mileage.

The notion of hyperbolicity for polynomials has been around since its introduction by L. Gårding sixty years ago [**Går51**].[3] The notion of stability for real or complex polynomials in several variables goes back at least to multivariate versions of Hurwitz stability in the 1980's, but appears to have become established as tools for combinatorialists only in the last five years [**COSW04, Brä07, BBL09, Gur08**]. These two related notions have now been applied in several seemingly unrelated fields.

Gårding's original purpose was to find the right condition to guarantee analytic stability of a PDE as it evolves in time from an original condition in space; the notion of stability here is that small perturbations of the initial conditions should produce small perturbations of the solution at time t. That Gårding found the right definition is undeniable because his result (Theorem 2.1 below) is an equivalence. His proof uses properties of hyperbolic functions and their cones that are derived directly from the definitions. Specfically, in order to construct solutions to boundary value problems via the **Riesz kernel**, an inverse Fourier transform is computed on a linear space $\{\mathbf{x} + i\mathbf{y} : \mathbf{y} \in \mathbb{R}^d\}$; hyperbolicity in the direction $\mathbf{x}$ is used to ensure that this space does not intersect the set $\{q = 0\}$ where q^{-a} is singular. Gårding developed some properties of hyperbolic polynomials in this paper and later devoted a separate paper to the further development of some inequalities [**Går59**].

Twenty years later, with Atiyah and Bott [**ABG70**], this work was extended considerably. They were able to compute inverse Fourier transforms more explicitly for a number of homogeneous hyperbolic functions. To do this, they developed semi-continuity properties of hyperbolic functions. These properties enable the construction of certain vector fields and deformations, which in turn are used to deform the plane $\{\mathbf{x} + i\mathbf{y} : \mathbf{y} \in \mathbb{R}^d\}$ into a cone on which the inverse Fourier transforms are directly integrable. Direct integrability provides useful estimates and also enables explicit computation via topological dimension reduction theorems of Leray and Petrovsky. My introduction to this exceptional body of work came in the early 2000's when

[3]Gårding credits many of the ideas to I. G. Petrovsky in his seminal paper [**Pet45**].

I needed to apply these deformations to multivariate analytic combinatorics; refining the relatively crude techniques in [**PW02, PW04**] led to [**BP11**].

In control theory, a continuous-time (respectively discrete-time) system whose transfer function is rational will be stable if its poles all lie in the open left half-plane (respectively the open unit disk). Accordingly, we say a univariate polynomial is **Hurwitz stable** (respectively **Schur stable**) if its zeros are all in the open left half-plane (respectively the open unit disk). Multivariate generalizations of Hurwitz and Schur stability have arisen in a variety of problems. The development of these generalizations splits into two veins. The review paper [**Sok05**] surveys a number of "hard" results on zero-free regions, meaning results that give good estimates on zero-free regions that depend on specific parameters of the models or graphs from which the polynomial is formed. Most relevant to the applications in Sections 6 and 7 are "soft" results, which give simple domains free of zeros (such as products of half planes), valid for all graphs or for wide classes of graphs.

The development of this vein of multivariate stable function theory took place mainly in the area of statistical physics and combinatorial extensions thereof. Multivariate Schur stability arises in the celebrated Lee–Yang Theorem for ferromagnetic Ising models [**YL52, LY52**], where it implies the absence of phase transitions at nonzero magnetic field. Likewise, multivariate Hurwitz stability arises in electrical circuit theory [**FB85, COSW04, Sok11**] and matroid generalizations thereof [**COSW04, Brä07, WW09**], and has applications to combinatorial enumeration [**Wag06, Wag08**]. It also arises in generalizations of the Lee–Yang Theorem [**LS81**] and in the Heilmann-Lieb Theorem for matching polynomials (also known as monomer-dimer models) [**HL72, COSW04**]. It should be noted that Hurwitz stability [**Hur96**], differs from the notion in Definition 1.2 in that the zero-free region is the right half-plane rather than the upper half-plane. For general complex polynomials the two notions are equivalent under a linear change of variables, but for real polynomials the two notions are very different. Except for brief portions of Sections 4 and 5, our concern will be with stability as defined in Definition 1.2

Multivariate stability behaves nicely under certain transformations. The closure properties were investigated further by Borcea and Brändén [**BB09a**]. These closure properties turn out to have powerful implications for systems of negatively dependent random variables, as worked out in 2009 by Borcea, Brändén and Liggett [**BBL09**]. My second brush with this subject was when I needed to apply some of their results to determinantal point processes [**PP11**].

These two encounters with hyperbolic/stable polynomials in seemingly dissimilar areas of mathematics prompted my search for a unified understanding. As pointed out at the end of the introduction of [**BBL09**], such a viewpoint was taken 25 years ago by Gian-Carlo Rota.

> "The one contribution of mine that I hope will be remembered has consisted in just pointing out that all sorts of problems of combinatorics can be viewed as problems of the locations of zeros of certain polynomials..."

to which we would add "and also problems in differential equations, number theory, probability, and perhaps many more areas."

Presently, the greatest interest in the subject of hyperbolic/stable polynomials seems to be its potential for unlocking some combinatorial conjectues concerning determinants such as the Bessis–Moussa–Villani conjecture, Lieb's "permanent on top" conjecture and extensions of the van der Waerden conjecture, proved in 1979, but still an active subject (see, e.g., [**Gur06**]). The distances between areas of mathematics in which hyperbolicity plays a role are such that it was difficult to find a primary subject classification for these lectures. It also poses a unique challenge for me, as I am an expert in none of these, save for applications to analytic combinatorics. I therefore apologize in advance for any historical inaccuracies perpetrated here, or for idiosyncratic viewpoints arising from my personal history with the subject. Here follows an attempt to lay out its two main pillars, hyperbolicity and stability, and to follow their progress up to the present day.

Part I: Hyperbolicity

2. Origins, definitions and properties

2.1. Relation to the propagation of wave-like equations. Gårding's objective was to prove stability results for wave-like partial differential equations. Endow the space of smooth complex valued functions of d real variables, denoted $C^\infty(\mathbb{R}^d)$, with the topology of uniform convergence on compact sets of all the partial derivatives. For any polynomial q in d variables, let $D[q] = q(\partial/\partial \mathbf{x})$ denote the linear differential operator obtained by formally substituting $(\partial/\partial x_i)$ for x_i in q. If we consider $\mathbb{R}^d$ as spacetime with the positive time direction given by a vector ξ, and denote by H_ξ the hyperplane orthogonal to ξ, then the boundary value equation

$$D[q](f) = 0, \qquad f = g \text{ on } H_\xi \tag{2.1}$$

on the halfspace $\{\mathbf{x} : \mathbf{x}\cdot\xi \geq 0\}$ may be interpreted as the evolution of g under the equation $D[q](f) = 0$ with initial condition g. One would expect $m - 1$ further initial conditions, such as $(\partial^j/\partial x_\xi^j)f = g_j$ for $1 \leq j \leq m-1$, where $(\partial/\partial x_\xi)$ denotes a derivative in direction ξ and m is the degree of q.

We consider (2.1) to be **stable under perturbations** (in time direction ξ) if for all sequences $\{f_n\}$ of functions in $C^\infty(\mathbb{R}^d)$ satisfying $D[q](f_n) = 0$, convergence to zero of the restrictions of f_n to H_ξ implies convergence to zero of f_n on the whole space. Intuitively, arbitrary small change in the initial conditions cannot lead to macroscopic changes in the evolution at a later time. The first of Gårding's results can be stated as follows.

THEOREM 2.1 ([**Går51**, Theorem III]). *The differential equation* (2.1) *is stable in direction ξ if and only if q is hyperbolic in direction ξ.*

It should be noted that the definition of hyperbolicity in direction $\mathbf{x}$ in Gårding's 1951 paper is that $q(t\mathbf{x} + i\mathbf{y}) \neq 0$ when $\mathbf{y} \in \mathbb{R}^d$ and $t > t_0$. This does not appear to me to be equivalent to Definition 1.1, which appears in all later work including [**ABG70**]. The two definitions specialize to the same thing when q is homogeneous. Although the primary focus of these notes is on later uses outside of PDE's, I will give indications of the proof. Not only does this satisfy curiosity and a historical sense, but it promotes the goal of using some of the geometric understanding when dealing with contemporary, more algebraic problems.

To keep things simple, assume q is homogeneous. Examples will be discussed shortly, but for now, a good mental picture is to keep in mind the example $d = 3$ and $q(x, y, z) = z^2 - x^2 - y^2$. Even though $\mathbb{R}^d$ and its dual are isomorphic, I find it useful to classify a vector as belonging to the dual space $(\mathbb{R}^d)^*$ rather than $\mathbb{R}^d$ if it plays the role of a linear functional on the original space. For example the frequency ξ of a wave $\mathbf{x} \mapsto \exp(i\xi \cdot \mathbf{x})$ is thought of as living in $(\mathbb{R}^d)^*$.

FORWARD DIRECTION: For any $\xi \in (\mathbb{R}^d)^*$, let f_ξ denote the function $\mathbf{x} \mapsto e^{i\xi \cdot \mathbf{x}}$ mapping $\mathbb{R}^d$ to $\mathbb{C}$. Of course $(\mathbb{R}^d)^*$ is isomorphic to $\mathbb{R}^d$ but we call it $(\mathbb{R}^d)^*$ to remind ourselves that ξ runs over frequency space. When $\mathbf{r}$ is real, the function $f_\mathbf{r}$ is a sinusoidal wave and is bounded, as in Figure 1. Let us assume without loss of generality that $\xi = (0, \ldots, 0, 1)$. If $q(\xi) = 0$ then f_ξ is a solution to $D[q](f) = 0$. If q is not hyperbolic in direction ξ then there is some $(r_1, \ldots, r_{d-1})$ such that the d solutions to $q(r_1, \ldots, r_{d-1}, t) = 0$ include at least one conjugate pair of values that are not real. Denoting by r_d the one with the negative imaginary part, we see that $f_\mathbf{r}(\mathbf{x})$ grows exponentially as x_d increases. By homogeneity, $f_{\lambda\mathbf{r}}$ grows at λ times the exponential rate. Choose c_λ so that $c_\lambda f_{\lambda\mathbf{r}}(0, \ldots, 0, 1) = 1$. Then $f_{n\mathbf{r}} \to 0$ on the hyperplane H_ξ in the topology of uniform convergence (due to periodicity, no compact set restriction is required), while $f_{n\mathbf{r}} \equiv 1$ at $(0, \ldots, 0, 1)$, proving the contrapositive, namely that lack of hyperbolicity implies lack of stability. □

BACKWARD DIRECTION, HANDWAVING ARGUMENT: For fixed $(r_1, \ldots, r_{d-1})$, the set $\{f_\mathbf{r} : \mathbf{r} \in (\mathbb{R}^d)^*\}$ of solutions to $D[q]f = 0$ spans a vector space $V(r_1, \ldots r_{d-1})$ of dimension k. Since we are waving our hands, we have assumed **strong** hyperbolicity, namely that the d values of r_d are distinct. Any integral $\int c(r_1, \ldots, r_{d-1}) dr_1 \cdots dr_{d-1}$ is also a solution, where $c(\cdot)$ is a section of $V(\cdot)$. The handwaving part is that this gives all solutions of $D[q]f = 0$. Assuming this, we write a generic solution f as such an integral. Each f_ξ evolves unitarily, so more handwaving along the lines of a Parseval relation shows that if f and its first k derivatives are small at time zero then the time t value is small as well. □

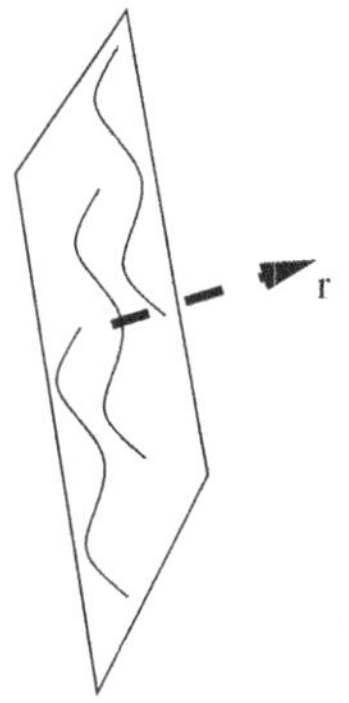

FIGURE 1. $f_\mathbf{r}$ when $\mathbf{r}$ is real.

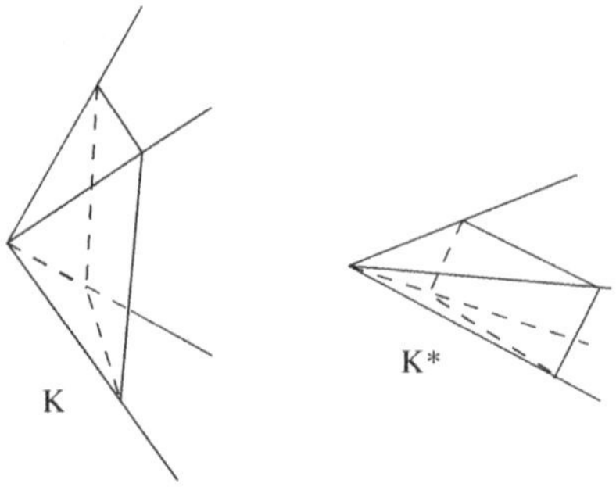

FIGURE 2. The cone K and its dual K^*.

Removing all the handwaving takes considerable work, beyond our scope here. The key is the construction of the Riesz kernel, which is interesting enough to merit a brief digression. Suppose that q is m-homogeneous and hyperbolic in direction $\mathbf{x}$ and let α is any complex number. Define the Riesz kernel Q_α by

$$Q_\alpha(\mathbf{y}) := (2\pi)^{-d} \int_{\mathbf{z} \in \mathbf{x} + i\mathbb{R}^d} q(\mathbf{z})^{-\alpha} \exp(\mathbf{y} \cdot \mathbf{z}) \, d\mathbf{z} \, . \tag{2.2}$$

We have not yet defined cones of hyperbolicity, but let us nonetheless try to visualize why the Riesz kernel is well defined and supported on the dual cone. To explain the terminology, the convex dual of a cone $K \subseteq \mathbb{R}^d$ is the cone $K^* \subseteq (\mathbb{R}^d)^*$ consisting of vectors $\mathbf{y}$ such that $\mathbf{w} \cdot \mathbf{y} \geq 0$ for all $\mathbf{w} \in K$. Duality maps fat cones to skinny cones and vice versa (see Figure 2). The connected components of the open set $\{\mathbf{w} \in \mathbb{R}^d : q(\mathbf{w}) \neq 0\}$ are cones. If convex they have duals. Let C be the component containing $\mathbf{x}$. We will see later that hyperbolicity guarantees this is one of the convex ones. Hyperbolicity in direction $\mathbf{x}$ guarantees that q is nonvanishing on the domain of integration $\mathbf{x} + i\mathbb{R}^d$. This is illustrated in Figure 3, where the dashed lines signify that the contour of integration is varying in imaginary directions only and therefore does not hit the surface $\{q = 0\}$. The magnitude of the exponential term is constant, whence the integral is convergent at infinity when $m \, \mathrm{Re}\,\{\alpha\} > 1 + d$. Because the chain of integration avoids $\{q = 0\}$, integrability is assured as is a well defined branch of $q^{-\alpha}$. It follows easily from Cauchy's Theorem that the Riesz kernel is independent of the

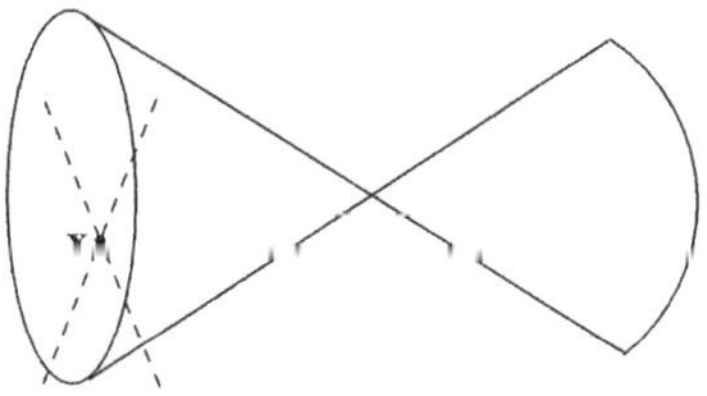

FIGURE 3. Chain of integration in the definition of Q_α.

choice of $\mathbf{x}$ within the component C. Furthermore, if $\mathbf{y} \in (\mathbb{R}^d)^*$ is a vector for which some $\mathbf{x} \in C$ has $\mathbf{x} \cdot \mathbf{y} < 0$, then it is clear that the integral defining $Q_\alpha(\mathbf{y})$ can be deformed so as to be arbitrarily small, hence is equal to zero. In other words, Q_α is supported on the set of $\mathbf{y}$ for which there is no such $\mathbf{x}$, which is precisely the dual cone C^*. □

To complete the sketchy proof of the backward direction, convolving with the Riesz kernel gives an operator I_q inverted by $D[q]$. This constuction is stable under perturbations. If $D[q]f = 0$, one shows that f is equal to $(I - I_q D[q])\tilde{f}$, where $\tilde{f}$ is defined stably in terms of f and its first $d-1$ normal derivatives on H_ξ. Small changes in the initial conditions thus give rise to small changes in $\tilde{f}$, which give rise to small changes in f, proving the theorem.

The Riesz kernels with parameters α and $\alpha - 1$ are related by a differential identity as long as $\alpha \neq 1$. When $\alpha = 1$, the Riesz kernel is called the **fundamental solution** by reason of a result of Gårding. This is explained more fully in Theorem 2.8 below, but the short statement, found for instance in [**Gül97**, Theorem 2.2], is that the solution of

$$D[q]f(\mathbf{x}) = \delta(\mathbf{x})$$

exists, is unique, and is supported on the dual cone, C^*.

2.2. Homogeneous hyperbolic polynomials. Many of the properties of hyperbolic polynomials are easier to state, prove and understand in the homogeneous case. Throughout this section, therefore, we deal only with homogeneous polynomials. We will use p rather than q as a visual cue when speaking about homogeneous polynomials. Hyperbolicity is easily seen to be equivalent to the equivalent to the following "real root" property.

PROPOSITION 2.2. *The homogeneous polynomial p is hyperbolic in direction $\mathbf{x}$ if and only if for any $\mathbf{y} \in \mathbb{R}^d$ the univariate polynomial $t \mapsto p(\mathbf{y} + t\mathbf{x})$ has only real roots.*

PROOF: Because p is homogeneous, when $\lambda \neq 0$, we have $p(\lambda \mathbf{z}) = 0$ if and only if $p(\mathbf{z}) = 0$. With $\lambda = is$ for some real nonzero value of s we see that $p(\mathbf{x} + i\mathbf{y}) \neq 0$ for all $\mathbf{y} \in \mathbb{R}^d$ (the definition of hyperbolicity) is equivalent to $p(\mathbf{y} + is\mathbf{x}) \neq 0$ for all $\mathbf{y} \in \mathbb{R}^d$ and all nonzero real s. This is equivalent to $p(\mathbf{y} + t\mathbf{x} + is\mathbf{x}) \neq 0$ for all $\mathbf{y} \in \mathbb{R}^d$ and real s, t with $s \neq 0$. Writing $z = t + is$, this is equivalent to $p(\mathbf{y} + z\mathbf{x}) \neq 0$ when z is not real. □

Denote by $t_k(\mathbf{x}, \mathbf{y})$ the roots of $t \mapsto p(\mathbf{y} + t\mathbf{x})$. Then

$$p(\mathbf{y} + t\mathbf{x}) = p(\mathbf{x}) \prod_{k=1}^{m} [t - t_k(\mathbf{x}, \mathbf{y})] . \tag{2.3}$$

By Proposition 2.2, when p is hyperbolic in direction ξ, all values $t_k(\mathbf{x}, \mathbf{y})$ are real, hence, setting $t = 0$, we see that $p/p(\mathbf{x})$ is a real polynomial. We

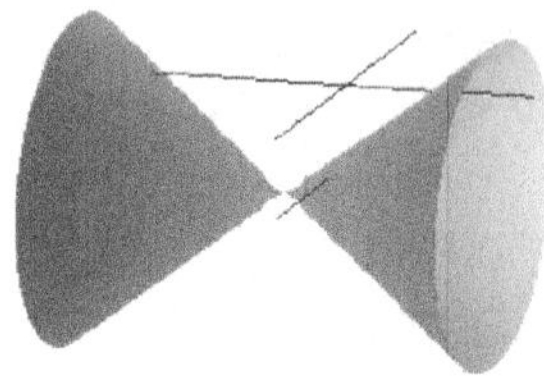

FIGURE 4. This Lorentzian quadratic is hyperbolic in the x-direction but not the y-direction.

see that little generality is lost in restricting our discussion of homogeneous polynomials to those with real coefficients.

The two nontrivial examples of hyperbolic polynomials given in Gårding's original paper, namely Lorentzian quadratics and the determinant, provide much of the intuition as to the meaning of hyperbolicity. We now discuss these. A preliminary observation is that, trivially, all real homogeneous polynomials of degree one are hyperbolic in all directions in which they do not vanish. Next, observe that hyperbolicity of the homogeneous polynomial p in direction ξ is preserved when p and ξ are transformed by the same invertible real linear map. This allows us to classify all nondegenerate quadratics: it suffices to consider all polynomials of the form $\sum_{j=1}^d \pm x_j^2$, hyperbolicity evidently being determined by signature. A necessary and sufficient condition for hyperbolicity is Lorentzian signature, that is precisely one sign different from the others.

EXAMPLE 2.3 (Lorentzian quadratics). *Let $p(\mathbf{x}) := x_1^2 - \sum_{j=2}^d x_j^2$. Real vectors ξ may be classified as time-like, light-like or space-like according to whether $p(\xi)$ is respectively positive, zero or negative. The time-like vectors form the open convex cone $\{\mathbf{x} : (x_2/x_1)^2 + \cdots + (x_d/x_1)^2 < 1\}$. Fix any time-like vector ξ. If $\eta = \lambda\xi$ then the line $t \mapsto \eta + t\xi$ contains a doubled root of p at the origin. For any other real η, the line $\eta + t\xi$ intersects the hyperplane H_ξ orthogonal to ξ at some point other than the origin and p takes a negative value there. On the other hand, the quadratic $p(\eta + t\xi)$ has positive leading term, so goes to $+\infty$ at $t = \pm\infty$. Hence the line intersects the zero set of p twice. The degree of p is two, hence for any η, the polynomial $t \mapsto p(\eta + t\xi)$ has all real roots. Hence p is hyperbolic. Replacing p by $-p$ does not affect hyperbolicity. Hence homogeneous quadratics with signature 1 or $d-1$ are hyperbolic in timelike directions.*

For the converse, elliptic quadratics have lines in every direction with no real roots, hence are obviously not hyperbolic. If $d > 4$ and there are at least two positive and two negative directions, the any line on which p has a minimum may be translated in a positive direction not on the line so that the minimum becomes positive, and similarly any line on which p has a maximum may be translated so that the maximum becomes negative. We conclude there are no directions of hyperbolicity.

The other classical example is as follows.

EXAMPLE 2.4 (Determinants of Hermitian matrices). *An $n \times n$ matrix M is Hermitian if $M_{ij} = \overline{M_{ji}}$. The space of Hermitian matrices of size n is parametrized by n^2 real parameters, these being the $\binom{n}{2}$ arbitrary complex numbers $\{M_{ij} : 1 \le i < j \le n\}$ and the n real numbers $\{M_{ii} : 1 \le i \le n\}$. Let p be the homogeneous polynomial on $\mathbb{R}^{n^2}$ defined by the determinant as a function of these n^2 real numbers. Hyperbolicity of p in the direction of any positive definite matrix A is equivalent to the well known fact that the zeros of $M + zA$ are all real if M is Hermitian and A is positive definite. To see this, let $z = a + bi$ with $b > 0$ and write*

$$\mathrm{Det}\,(M + zA) = \mathrm{Det}\,(bA)\mathrm{Det}\,\left[(bA)^{-1/2}(aA + M)(bA)^{-1/2} + iI\right],$$

where $(bA)^{1/2}$ is the Hermitian positive definite square root of bA. If this quantity is equal to zero then $-i$ is an eigenvalue of $(bA)^{-1/2}(aA+M)(bA)^{-1/2}$ which is impossible because this matrix is Hermitian.

There are several ways to construct new hyperbolic functions from old ones. Note that the first does not require homogeneity.

PROPOSITION 2.5 (products and polarization).

(i) Let q_1 and q_2 be hyperbolic with respect to $\mathbf{x}$. Then $q_1 q_2$ is also hyperbolic with respect to $\mathbf{x}$.

(ii) Let p be homogeneous of degree m and hyperbolic with respect to $\mathbf{x}$. Let $p_0, \ldots, p_m$ be the coefficients of $p(\mathbf{y} + t\mathbf{x})$ as a polynomial in t:

$$p(\mathbf{y} + t\mathbf{x}) = \sum_{k=0}^{m} p_k(\mathbf{y}) t^k . \tag{2.4}$$

Then p_k is hyperbolic with respect to $\mathbf{x}$ for all $0 \le k \le m$.

PROOF: The fact for products is immediate from the definition. For polarization, we first observe that hyperbolicity of p in direction $\mathbf{x}$ implies hyperbolicity of the directional derivative $D[\mathbf{x}][p]$ in direction $\mathbf{x}$. To see this, observe that $f(t)$ having only real roots implies $f'(t)$ has only real roots. Setting $f(t) = p(\mathbf{y} + t\mathbf{x})$, we see that $f'(t)$, namely $t \mapsto D_{\mathbf{x}}(p)(\mathbf{y} + t\mathbf{x})$, has only real roots for any $\mathbf{y} \in \mathbb{R}^d$, which is hyperbolicity of $D_{\mathbf{x}}(p)$ in direction $\mathbf{x}$.

Now consider the expansion (2.4) of $p(\mathbf{y} + t\mathbf{x})$ into homogeneous parts. The parts are given by

$$p_k(\mathbf{y}) = \frac{1}{k!} \left(\frac{d}{dt}\right)^k \Big|_{t=0} p(\mathbf{y} + t\mathbf{x}) = (D_{\mathbf{x}})^k (p)(\mathbf{y}) .$$

Hyperbolicity in direction $\mathbf{x}$ is stable under $D_{\mathbf{x}}$, which shows that p_k is hyperbolic in direction $\mathbf{x}$. □

2.3. Cones of hyperbolicity for homogeneous polynomials. Hyperbolic polynomials have associated with them certain convex cones. In the homogeneous case the definition is simple and is contained in the following proposition. This was proved first in [**Går51**] and reproduced many times; the proof below follows [**Gül97**]. Let p be a (complex) homogeneous hyperbolic polynomial, hyperbolic in direction ξ. Dividing by a real multiple of $p(\xi)$ we may assume, cf. (2.3), that p is real and $p(\xi) = 1$.

PROPOSITION 2.6. *Let $K(p, \xi)$ denote the connected component of the set $\mathbb{R}^d \setminus \{\mathbf{x} : p(\mathbf{x}) = 0\}$ that contains ξ.*

(i) p is hyperbolic in direction $\mathbf{x}$ for every $\mathbf{x} \in K(p, \xi)$.
(ii) The set $K(p, \xi)$ is an open convex cone; we call this a ***cone of hyperbolicity*** *for p.*
(iii) $K(p, \xi)$ is equal to the set K of vectors $\mathbf{x}$ for which all roots of $t \mapsto p(\mathbf{x} + t\xi)$ are real and negative, which (by hyperbolicity of p in direction ξ) is the same as the set of vectors $\mathbf{x}$ for which no root of this polynomial is real and nonnegative.

PROOF: By continuity of the roots of a polynomial with respect to its coefficients, the set K is open. Also $\xi \in K$ because $p(\xi + t\xi) = (1+t)^m p(\xi)$ where m is the degree of the homogeneous polynomial p. If $\mathbf{x}$ is in the closure $\overline{K}$ then, again by continuity of the roots, $p(\mathbf{x} + t\xi) \neq 0$ for $t > 0$; in particular, $p(\mathbf{x}) \neq 0$ implies $\mathbf{x} \in K$, which means that K is closed in $\mathbb{R}^d \setminus \{p = 0\}$. Also K is connected: by definition of K, if $\mathbf{x} \in K$ then $\mathbf{x} + t\xi \in K$ for $t > 0$, and by homogeneity $s\mathbf{x} + t\xi \in K$ for $s, t > 0$, and sending s to zero proves that K is star-convex at ξ (hence connected). Being connected and both closed and open in $\mathbb{R}^d \setminus \{p = 0\}$, K is a component of $\mathbb{R}^d \setminus \{p = 0\}$ and is thus equal to $K(p, \xi)$.

Next we check that for $\mathbf{v} \in K$, $\mathbf{x}$ real, and s and t complex,

$$p(\mathbf{x} + t\mathbf{v} + s\xi) \neq 0 \text{ whenever } \Im(s) \leq 0, \Im(t) \leq 0, \Im(s) + \Im(t) < 0\,. \tag{2.5}$$

First, suppose $\Im(s) < 0$. If $u \geq 1$ then the polynomial

$$t \mapsto p\left[\frac{\mathbf{x}}{u} + t\mathbf{v} + \left(\frac{s}{u} + i\left(\frac{1}{u} - 1\right)\right)\xi\right] \tag{2.6}$$

is nonvanishing for real t because this is p evaluated at the sum of a real vector and a complex multiple of ξ with nonzero imaginary part. As $u \to \infty$, the number of roots of (2.6) in the lower half-plane remains constant. The limit at $u = \infty$ is $t \mapsto p(t\mathbf{v} - i\xi)$. This has roots in the upper half-plane because its roots are $-i$ divided by the roots of $t \mapsto p(\mathbf{v} + t\xi)$, the latter of which we have seen to be negative real. We conclude that for $u = 1$ there are no roots in the lower half-plane, in other words,

$$p(\mathbf{x} + t\mathbf{v} + s\xi) \neq 0 \text{ whenever } \Im(s) < 0, \Im(t) \leq 0\,. \tag{2.7}$$

To see that this also holds for $\Im(s) = 0$ and $\Im(t) < 0$, note that K is open, so $\mathbf{v} - \epsilon\xi \in K$ for some $\epsilon > 0$, whence

$$p(\mathbf{x} + t\mathbf{v} + s\xi) = p(\mathbf{x} + t(\mathbf{v} - \epsilon\xi) + (s + \epsilon t)\xi)$$

which is nonzero by (2.7). This completes the verification of (2.5).

Setting $s = 0$ in (2.5) shows that any root of $t \mapsto p(\mathbf{x}+t\mathbf{v})$ satisfies $\Im(t) \geq 0$. Because p is real, complex conjugation may be applied to the entire argument, showing that also $\Im(t) \leq 0$, and hence that all roots of $t \mapsto p(\mathbf{x} + t\mathbf{v})$ are real. The vector $\mathbf{v}$ was chosen arbitrarily in K. We conclude that p is hyperbolic with respect to every $\mathbf{v} \in K$. It follows that K is star convex with respect to $\mathbf{v}$ as well, and since $\mathbf{v} \in K$ is arbitrary, that K is convex. □

Because the cones of hyperbolicity are characterized as components of the nonzero set in real space, it follows that the cones of hyperbolicity of pq are pairwise intersections of the cones of hyperbolicity of p with the cones of hyperbolicity of q.

EXAMPLE 2.7 (coordinate planes). *Each coordinate function z_j is homogeneous of degree 1, therefore hyperbolic in every direction not contained in the plane $\{z_j = 0\}$. The cones of hyperbolicity are the two half spaces bounded by this plane. It follows that the product $\prod_{j=1}^d z_j$ is hyperbolic in every direction in which no coordinate vanishes, and that the cones of hyperbolicity are the orthants. This is also obvious from Figure 5.*

Returning briefly to the origins of hyperbolicity theory in the properties of wave-like partial differential equations, the following result was proved in [**ABG70**, Theorem 4.1]. It says roughly that cones of hyperbolicity are propagation cones for solutions to wave-like equations. Given any convex cone $K \subseteq \mathbb{R}^d$ over the origin, let K^* denote the dual cone of vectors $\mathbf{v}^* \in (\mathbb{R}^d)^*$ such that $\langle \mathbf{v}^*, \mathbf{x} \rangle \leq 0$ for all $\mathbf{x} \in K$. Recall the linear partial differential operator $D[p]$ defined by substituting $\partial/\partial x_i$ for x_i in p. Let $E = E(p, \mathbf{x})$ be a solution to

$$D[p]E = \delta_{\mathbf{x}} \tag{2.8}$$

where $\delta_{\mathbf{x}}$ is a delta function at $\mathbf{x}$.

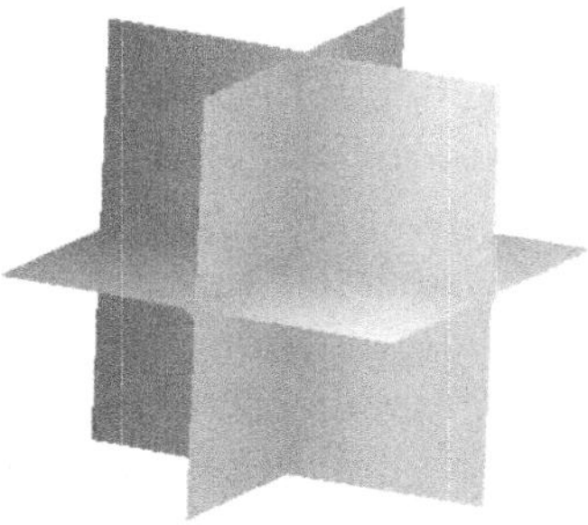

FIGURE 5. The cones of hyperbolicity for $\prod_{j=1}^d z_j$ are the orthants.

THEOREM 2.8. *Suppose the homogeneous polynomial p is hyperbolic in direction ξ. Then there exists a unique distribution E solving* (2.8) *and its support is the cone $K(p,\xi)^*$ dual to the cone of hyperbolicity of p containing ξ. In fact E is given as in inverse Fourier transform:*

$$E(p,\xi) = (2\pi)^{-d} \int_{\mathbb{R}^d} p_-(\mathbf{y})^{-1} e^{i\langle \mathbf{x},\mathbf{y}\rangle}\, d\mathbf{y}$$

where $p_-(y)^{-1} := \lim_{t\downarrow 0} p(\mathbf{y} - it\nu)^{-1}$ *as distributions and ν is any element of $K(p,\xi)$.* □

Although there is no time here to explore the properties of hyperbolic polynomials as barrier functions, I will mention this example briefly, as it has proved to be of some importance to convex programming.

EXAMPLE 2.9 (self-concordant barrier functions). *Let $Q \subseteq \mathbb{R}^d$ be an open convex set. A function $F : Q \to \mathbb{R}$ is a* ***logarithmically homogeneous self-concordant barrier function*** *for Q if it is smooth, convex, and satisfies several properties:*

$$F(\mathbf{x}) \to \infty \text{ as } \mathbf{x} \to \partial Q$$

$$\begin{aligned} \left|D^3F(\mathbf{x})[u,u,u]\right| &\leq 2\left(D^2F(\mathbf{x})[u,u]\right)^{3/2} \\ ||DF(\mathbf{x})[u]|^2 &\leq \theta D^2F(\mathbf{x})[u,u] \\ F(t\mathbf{x}) &= F(\mathbf{x}) - \theta \log t. \end{aligned}$$

The interior point method for convex programming problems (see, e.g. [**NN93**]*) is based on finding such a function for a given region, Q. There is a universal construction but it is not always useful for computations. Also, properties beyond those satisfied by the universal construction are required for long term stability of the interior point method. Such properties will depend on the region Q. When Q is the cone of hyperbolicity for a homogeneous polynomial p, it turns out that the function $F(\mathbf{x}) := -\log p(\mathbf{x})$ is a logarithmically homogeneous self-concordant barrier function with a number of other useful properties. These are detailed in* [**Gül97**].

Hyperbolicity without homogeneity. Delving into the relation between hyperbolicity in the homogeneous and non-homogeneous cases, (definitions both given in Definition 1.1, we begin by looking at the homogeneous parts of a polynomial, f. Sorting the terms of f by total degree, the highest degree part will be denoted $LT(f)$ for "leading term(s)". The lowest degree is called the **localization** and will be discussed further in the next section. If f is hyperbolic, then both the leading term and the localization are hyperbolic as well (respectively [**ABG70**, Lemma 3.20] and [**ABG70**, Lemma 3.42]). By definition, the cones of hyperbolicity of a non-homogeneous polynomial f are the cones of hyperbolicity of $LT(f)$. When f is hyperbolic, these cones still characterize the directions of stable propagation of PDE's and supports of solutions to these.

Because hyperbolicity is easier to understand in the homogeneous case, Gårding looked for a converse to these and came up with a criterion for homogeneous polynomials called strong hyperbolicity. This is most naturally stated using the equivalent definition of hyperbolicity from Proposition 2.2: the roots of $p(\mathbf{y} + t\mathbf{x})$ must not only be real but also distinct (except when $\mathbf{y}$ is a multiple of $\mathbf{x}$, when all the zeros coincide perforce). We then have:

THEOREM 2.10. *If A is strongly hyperbolic then f is hyperbolic for any f such that $LT(f) = A$.*

This is proved as (3.9) of [**ABG70**] and in fact is already in [**Går51**] after the statement and proof of Lemma 2.5. Necessary and sufficient conditions for f to be hyperbolic are given in [**Hör83**, Theorem 12.4.6]; these were conjectured by Gårding and proved first by Svensson [**Sve68**].

3. Semi-continuity and Morse deformations

3.1. Localization. Given a function f analytic on a neighborhood of the origin, its order of vanishing is the least total degree of a nonvanishing term in its Taylor series. The sum of all such terms is called the homogeneous part of f at the origin and denoted $\mathsf{hom}(f)$. For any $\mathbf{x}$ we let

$$\mathsf{hom}(f, \mathbf{x}) := \mathsf{hom}(f(\mathbf{x} + \cdot))$$

denote the homogeneous part of f at $\mathbf{x}$. When $\mathbf{x} \neq 0$, Atiyah *et al.* [**ABG70**] define this by taking the $t = 0$ term of $t^m f(t^{-1}\mathbf{x}+\mathbf{y})$ as a function of $\mathbf{y}$, where $m - \deg(f)$, and they refer to this as the **localization** of f at $\mathbf{x}$.

PROPOSITION 3.1. *Let p be any hyperbolic homogeneous polynomial, and let m be its degree. Fix $\mathbf{x}$ with $p(\mathbf{x}) = 0$ and let $\tilde{p} := \mathsf{hom}(p, \mathbf{x})$ denote the leading homogeneous part of p at $\mathbf{x}$. If p is hyperbolic in direction $\mathbf{u}$ then $\tilde{p}$ is also hyperbolic in direction $\mathbf{u}$. Consequently, if B is any cone of hyperbolicity for p then there is some cone of hyperbolicity for $\tilde{p}$ containing B.*

PROOF: This follows from the conclusion (3.45) of [**ABG70**, Lemma 3.42]. Because the development there is long and complicated, we give here a short, self-contained proof, provided by J. Borcea [**BP11**, Proposition 2.8]. If Q is a polynomial whose degree at zero is k, we may recover its leading homogeneous part $\mathsf{hom}(Q)$ by

$$\mathsf{hom}(Q)(\mathbf{y}) = \lim_{\lambda \to \infty} \lambda^k Q(\lambda^{-1}\mathbf{y}) .$$

The limit is uniform as $\mathbf{y}$ varies over compact sets. Indeed, monomials of degree k are invariant under the scaling on the right-hand side, while monomials of degree $k + j$ scale by λ^{-j}, uniformly over compact sets.

Apply this with $Q(\cdot) = p(\mathbf{x} + \cdot)$ and $\mathbf{y} + t\mathbf{u}$ in place of $\mathbf{y}$ to see that for fixed $\mathbf{x}, \mathbf{y}$ and $\mathbf{u}$,

$$\tilde{p}(\mathbf{y} + t\mathbf{u}) = \lim_{\lambda \to \infty} \lambda^k p(\mathbf{x} + \lambda^{-1}(\mathbf{y} + t\mathbf{u}))$$

uniformly as t varies over compact sub-intervals of $\mathbb{R}$. Because p is hyperbolic in direction $\mathbf{u}$, for any fixed λ, all the zeros of this polynomial in t are real. Hurwitz' theorem on the continuity of zeros [**Con78**, Corollary 2.6] says that a limit, uniformly on bounded intervals, of polynomials having all real zeros will either have all real zeros or vanish identically. The limit $\tilde{p}(\mathbf{y}+t\mathbf{u})$ has degree $k \geq 1$; it does not vanish identically and therefore it has all real zeros. This shows $\tilde{p}$ to be hyperbolic in direction $\mathbf{u}$. □

DEFINITION 3.2 (family of cones in the homogeneous case). *Let p be a hyperbolic homogeneous polynomial and let B be a cone of hyperbolicity for p. If $p(\mathbf{x}) = 0$, define*

$$\mathbf{K}^{p,B}(\mathbf{x})$$

to be the cone of hyperbolicity of $\mathsf{hom}(p, \mathbf{x})$ *containing B, whose existence we have just proved. If $p(\mathbf{x}) \neq 0$ we define $\mathbf{K}^{p,B}(\mathbf{x})$ to be all of $\mathbb{R}^d$.*

EXAMPLE 3.3 (strongly hyperbolic functions). *As mentioned in passing in Section 2, a homogeneous polynomial p is said to be strongly hyperbolic if the roots of $p(\mathbf{y} + t\mathbf{x})$, in addition to being real, are distinct. Equivalently, the projective variety $\mathcal{V} := \{\mathbf{x} : q(\mathbf{x}) = 0\}$ is smooth. In this case, for each nonzero $\mathbf{x}$, the localization* $\mathsf{hom}(p, \mathbf{x})$ *has degree one. The cones of hyperbolicity $K^{p,B}(\mathbf{x})$ are halfspaces whose common tangent hyperplane is tangent to B at $\mathbf{x}$; the convexity of B implies that the tangent hyperplane is a support hyperplane to B at $\mathbf{x}$, and we see that indeed $K^{p,B}(\mathbf{x})$ contains B.*

Suppose $\mathbf{x}_n \to \mathbf{x}$. It is not in general true that $\mathsf{hom}(f, \mathbf{x}_n) \to \mathsf{hom}(f, \mathbf{x})$. However, it is true that

$$\mathsf{hom}(f, \mathbf{x}_n) = \mathsf{hom}(\mathsf{hom}(f, \mathbf{x}), \mathbf{x}_n)$$

for n sufficiently large. This implies that if p is homogeneous and hyperbolic with cone of hyperbolicity B, then $K^{p,B}(\mathbf{x})$ is semi-continuous in $\mathbf{x}$:

$$K^{p,B}(\mathbf{x}) \subseteq \liminf K^{p,B}(\mathbf{x}_n) \text{ as } \mathbf{x}_n \to \mathbf{x}\,. \tag{3.1}$$

This is proved in [**ABG70**, Lemma 5.9]. We will want a version of this valid for polynomials that are not necessarily homogeneous. If q is hyperbolic but not homogeneous, the cone $K^{q,B}(\mathbf{x})$ has not been defined. To do this, at least for some points $\mathbf{x}$, we need the notion of the amoeba of q.

3.2. Amoeba boundaries.

DEFINITION 3.4 (amoeba). *The* ***amoeba*** *of any polynomial q is defined to be the image of the zero set $\mathcal{V} := \{\mathbf{z} : q = 0\}$ of q under the coordinatewise log modulus map* $\text{ReLog}(\mathbf{z}) := (\log|z_1|, \ldots, \log|z_d|)$. *We denote this image by* $\mathsf{amoeba}(q)$.

The connected components of $\mathbb{R}^d \setminus \mathsf{amoeba}(q)$ are convex sets and are in one to one correspondence with the Laurent series expansions of F, with each expansion converging on precisely the set $\{\exp(\mathbf{x}+i\mathbf{y}) : \mathbf{y} \in B\}$ for some component B of $\mathbb{R}^d \setminus \mathsf{amoeba}(q)$. This is well known and is presented, for

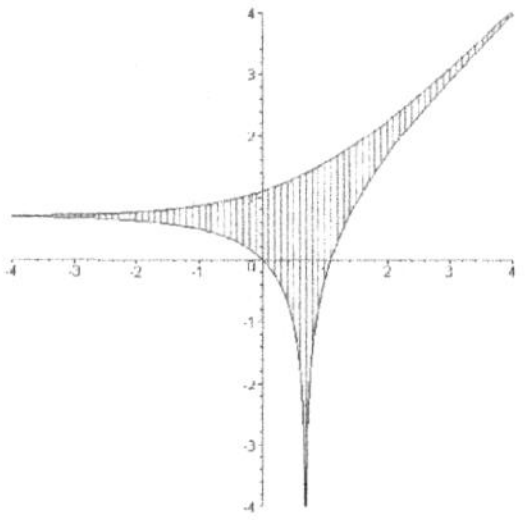

FIGURE 6. The amoeba of the polynoimal $2 - x - y$.

instance, in Chapter 6 of [**GKZ94**]. For those who have not seen an amoeba before, one of the simplest nontrivial amoebas is shown in Figure 6.

The following result, proved in [**BP11**, Proposition 2.12], defines a family of cones via localizations on the boundary of an amoeba.

PROPOSITION 3.5. *Let q be any polynomial and let B be a component of the complement of* $\mathsf{amoeba}(q)$. *Denote the boundary of B by ∂B. Fix $\mathbf{x} \in \partial B$ and let $f = q \circ \exp$ so that f vanishes at some point $\mathbf{x} + i\mathbf{y}$. Let $f_{\mathbf{y}} :=$* $\mathsf{hom}(f, \mathbf{x} + i\mathbf{y})$. *Then each $f_{\mathbf{y}}$ is hyperbolic (meaning hyperbolic in at least one direction) and one of its cones of hyperbolicity contains B. We denote this cone by $K^{q,B}(\mathbf{y})$. (The point $\mathbf{x}$ is considered fixed and is suppressed from the notation.)* □

One may extend semi-continuity for localizations beyond the homogeneous case, to the families in Proposition 3.5.

THEOREM 3.6 ([**BP11**, Corollary 2.15]). *If B is a component of* $\mathsf{amoeba}(q)^c$ *and $\mathbf{x} \in \partial B$, then the family of cones $K^{q,B}(\mathbf{y})$ defined in Proposition 3.5 satisfies*

$$K^{q,B}(\mathbf{y}) \subseteq \liminf K^{q,B}(\mathbf{y}_n) \text{ whenever } \mathbf{y}_n \to \mathbf{y}\,. \tag{3.2}$$

□

The fact that localizations of *any* polynomial to points on the amoeba boundary are hyperbolic allows us to give numerous examples fo hyperbolic polynomials beyond the classic ones: planes, quadrics, and the determinant function. For example, the homogeneous polynomial variety shown in Figure 7 is a localization of the famous so-called fortress generating function denominator.

It is instructive to see what role hyperbolicity plays in Theorem 3.6 by considering the counterexample $f(x, y, z) = xy + z^3$ as in Figure 8. As $\mathbf{y}$ varies over a neighborhood U of the origin, is it possible to choose open convex cones $\{K(\mathbf{y}) : \mathbf{y} \in U\}$ over $\mathbf{y}$ in such a way that $K(\mathbf{y})$ varies semi-continuously with $\mathbf{y}$ and each $K(\mathbf{y})$ is, locally, a subset of $\{\mathbf{y} : f(\mathbf{y}) \neq 0\}$? The points $\mathbf{y} = (0, y, 0)$ are forced to choose whether $K(\mathbf{y})$ contains points with positive $\mathbf{x}$ components or negative $\mathbf{x}$ components. One of these violates

FIGURE 7. Another hyperbolic homogeneous polynomial.

semi-continuity with cones $K(0,y,\epsilon)$ as $\epsilon \downarrow 0$ while the other violates semi-continuity with cones $K(0,y,-\epsilon)$ as $-\epsilon \uparrow 0$. The key to avoiding this in Theorem 3.6 is that $f_{\mathbf{y}}$ must be hyperbolic, and this cannot happen with hyperbolic functions.

3.3. Morse deformations. Suppose that $K(\mathbf{y})$ is any family of convex cones in $\mathbb{R}^d$ varying semi-conintuously with $\mathbf{y}$ in the sense of (3.2). Let $\mathbf{r}$ be any vector such that each $K(\mathbf{y})$ contains a vector $\mathbf{v}$ with $\mathbf{r} \cdot \mathbf{v} > 0$.

PROPOSITION 3.7. *Over any compact set S a section $\{\mathbf{v}(\mathbf{y}) : \mathbf{y} \in S\}$ may be chosen continuously with $\mathbf{v}(\mathbf{y}) > 0$ for all $\mathbf{y} \in S$.*

PROOF: Given $\mathbf{y}$, we may choose $\mathbf{v}$ in the interior of $K(\mathbf{y})$ with $\mathbf{v} \cdot \mathbf{r} > 0$, whence by semi-continuity, $\mathbf{v} \in K(\mathbf{y}')$ for all $\mathbf{y}'$ in some neighborhood of $\mathbf{y}$ which we denote $\mathcal{N}(\mathbf{y}, \mathbf{v})$. We may cover the compact set S with finitely many of these neighborhoods, $\{\mathcal{N}(\mathbf{y}^{(k)}, \mathbf{v}^{(k)}) : 1 \leq k \leq m\}$. Let $\{\psi^{(k)} : 1 \leq k \leq m\}$ be a partition of unity subordinate to this cover and for any

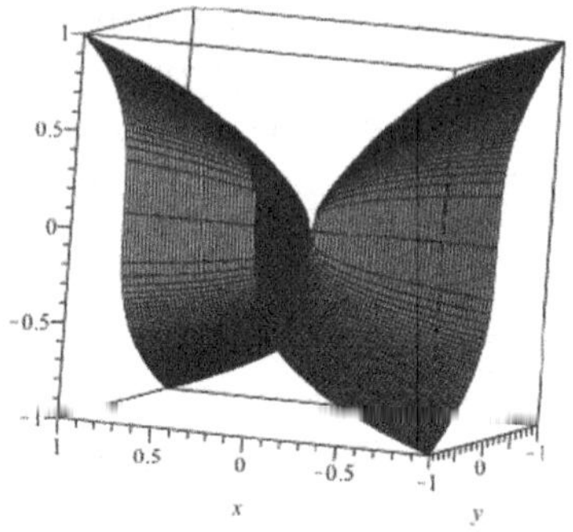

FIGURE 8. The sheared cubic $xy + z^3$.

$\mathbf{y} \in S$ define

$$\mathbf{v}(\mathbf{y}) := \sum_{k=1}^{m} \psi^{(k)}(\mathbf{y})\mathbf{v}^{(k)} .$$

It is clear that $\mathbf{r} \cdot \mathbf{v}(\mathbf{y}) > 0$ for all $\mathbf{y} \in S$. For each $\mathbf{y} \in S$, if $\psi^{(k)}(\mathbf{y}) > 0$ then $\mathbf{v}^{(k)} \in K(\mathbf{y})$. By convexity of $K(\mathbf{y})$, it follows that $\mathbf{v}(\mathbf{y}) \in K(\mathbf{y})$. □

Two applications of this are as follows. Suppose p is a homogeneous hyperbolic function and let B be a cone of hyperbolicity for p. Due to homogeneity, $K^{p,B}(\lambda\mathbf{y}) = K^{p,B}(\mathbf{y})$, whence the family may be described as $\{K^{p,B}(\hat{\mathbf{y}}) : \hat{\mathbf{y}} \in S^{d-1}\}$. Suppose $\mathbf{r}$ is such that each $K(\mathbf{y})$ contains a $\mathbf{v}$ with $\mathbf{r} \cdot \mathbf{v} > 0$. Because S^{d-1} is compact, an application of Proposition 3.7 yields a continuous section $\{\mathbf{v}(\mathbf{y}) : \mathbf{y} \in S^{d-1}\}$ with $\mathbf{v}(\mathbf{y}) \in K^{p,B}(\mathbf{y})$ and $\mathbf{r} \cdot \mathbf{v}(\mathbf{y}) > 0$. Extending to all of $\mathbb{R}^d \setminus \{\mathbf{0}\}$ by $\mathbf{v}(\lambda\mathbf{y}) := \lambda\mathbf{v}(\mathbf{y})$, we arrive at:

COROLLARY 3.8. *There is a vector field $\mathbf{v}(\cdot)$ on $\mathbb{R}^d$ which is 1-homogeneous, vanishes only at $\mathbf{0}$, is a section of $K^{p,B}(\cdot)$, and for some $\epsilon > 0$ satisfies $\mathbf{r} \cdot \mathbf{v}(\mathbf{y}) \geq \epsilon|\mathbf{y}|$.* □

EXAMPLE 3.9. *If the homogeneous polynomial q is strongly hyperbolic then its zero set is the cone over a smooth projective hypersurface and each $K(\mathbf{y})$ is a halfspace (see Example 3.3). The condition that every $K(\mathbf{y})$ contain a vector $\mathbf{v}$ with $\mathbf{r} \cdot \mathbf{v} > 0$ is the same as requiring that $\mathbf{r}$ not be the outward normal to the bounding hyperplane of $K(\mathbf{y})$. In other words, $\mathbf{r}$ may not be on the boundary of the dual cone to B.*

The second application is when B is a component of $\mathsf{amoeba}(q)^c$. In this case the family $\{f_{\mathbf{y}}\}$ is periodic with period 2π in each coordinate. Accordingly, the cones $K(\mathbf{y}) = K^{q,B}(\mathbf{y})$ from Proposition 3.5 are indexed by a compact set, namely the torus $\mathbb{R}^d/(2\pi\mathbb{Z})^d$. Applying Proposition 3.7 then yields:

COROLLARY 3.10. *If each cone $K^{q,B}(\mathbf{y})$ in Proposition 3.5 contains a vector $\mathbf{v}$ with $\mathbf{r} \cdot \mathbf{v} > 0$ then there is a continuously varying section $\{\mathbf{v}(\mathbf{y}) : \mathbf{y} \in \mathbb{R}^d\}$ with $\mathbf{v}(\mathbf{y}) \in K^{q,B}(\mathbf{y})$ and $\mathbf{r} \cdot \mathbf{v}(\mathbf{y}) > 0$ for all $\mathbf{y}$.* □

With these vector fields in hand we have enough to carry out the programs of [**ABG70**] and [**BP11**]. I will briefly describe the former, then go into a little more detail on the latter. Let q be a homogeneous polynomial, strongly hyperbolic in the direction $\mathbf{x}$. We wish to compute its inverse Fourier transform, which we have called the Riesz kernel, Q_α. We recall that its value at $\mathbf{y}$ is given by

$$Q_\alpha(\mathbf{y}) := (2\pi)^{-d} \int_{\mathbf{z}\in\mathbf{x}+i\mathbb{R}^d} q(\mathbf{z})^{-\alpha} \exp(\mathbf{y} \cdot \mathbf{z})\, d\mathbf{z} . \qquad (2.2)$$

We set $\mathbf{r} = -\mathbf{y}$ and assume $\mathbf{r}$ not to be on the boundary of B^*, the dual cone of the cone B of hyperbolicity of q that contains $\mathbf{x}$. As we have seen in Example 3.9, this guarantees the existence of the 1-homogeneous vector field $\mathbf{v}(\cdot)$.

The sets $K(\mathbf{y})$ are convex and each contains both $\mathbf{x}$ and $\mathbf{v}(\mathbf{y})$. Therefore, each contains the line segment joining $\mathbf{x}$ to $\mathbf{v}(\mathbf{y})$. Define a homotopy by

$$H(\mathbf{u}, t) := i\mathbf{u} + (1-t)\mathbf{x} + t\mathbf{v}(\mathbf{u}) .$$

This deforms the domain of integration $\mathbf{x} + i\mathbb{R}^d$ in (2.2) to the cone $\mathcal{C} := \{i\mathbf{u} + \mathbf{v}(\mathbf{u}) : \mathbf{u} \in \mathbb{R}^d\}$ while avoiding the set $\mathcal{V}$ where q vanishes. Cauchy's theorem implies that deforming the contour does not change the integral and therefore that

$$Q_\alpha(\mathbf{y}) = \int_{\mathcal{C}} q(\mathbf{z})^{-\alpha} \exp(\mathbf{y} \cdot \mathbf{z})\, d\mathbf{z} .$$

Omitting pages of detail and skipping to the punch line, this representation allows us to factor the integral. Integrating radially reduces the integral to an integral over the **leray cycle**, which is a $(d-1)$-dimensional homology class in the complement of $\{q = 0\}$ in $\mathbb{CP}^{d-1}$. The dimension reduction allows a number of explicit computations which are carried out in [**ABG70**, Section 7].

3.4. Asymptotics of Taylor coefficients. Let $F = P/Q = \sum_{\mathbf{r}} a_{\mathbf{r}} \mathbf{Z}^{\mathbf{r}}$ be a Laurent series expansion converging on the component B of $\mathsf{amoeba}(Q)^c$. The coefficients $\{a_{\mathbf{r}}\}$ of F may be recovered via Cauchy's formula

$$a_{\mathbf{r}} = (2\pi i)^{-d} \int_T \mathbf{Z}^{-\mathbf{r}} F(\mathbf{Z}) \frac{d\mathbf{Z}}{\mathbf{Z}} \tag{3.3}$$

where T is the torus $T(\mathbf{x}) = \{\exp(\mathbf{x} + i\mathbf{y}) : \mathbf{y} \in \mathbb{R}^d/(2\pi\mathbb{Z})^d\}$ for some $\mathbf{x} \in B$ and $d\mathbf{Z}/\mathbf{Z} := dz_1 \wedge \cdots \wedge dz_d/(z_1 \cdots z_d)$. The motivation for the evaluation or asymptotic estimation of the coefficients $\{a_{\mathbf{r}}\}$ comes from analytic combinatorics, where the primary object of study is the array $\{a_{\mathbf{r}}\}$, which counts something of interest. One constructs the generating function $F(\mathbf{Z}) := \sum_{\mathbf{r}} a_{\mathbf{r}} \mathbf{Z}^{\mathbf{r}}$ and hopes to identify a closed form representation of F. If $\{a_{\mathbf{r}}\}$ satisfy a recursion, then (depending on boundary values) this will usually succeed. A number of examples of combinatorial interest are surveyed in [**PW08**]. In most of the cases surveyed there, the generating function F is rational. In [**RW08**], based on a result of [**Saf00**], it is shown how to embed any algebraic function of d variables as a diagonal of a rational function of $(d+1)$ variables. Thus the evaluation of Taylor coefficients of rational functions solves the enumeration problem for any array of numbers whose generating function is rational or algebraic.

Exact evaluation of the Cauchy integral (3.3) is not easy but some methods are known for evaluating it asymptotically. In the case where the pole variety $\mathcal{V} := \{\mathbf{z} : Q(\mathbf{z}) = 0\}$ is smooth, a formula was given in [**PW02**]; a coordinate free version is given in [**BBBP08**]. Normal self-intersection in $\mathcal{V}$ can also be dealt with [**PW04**]. A number of examples from combinatorics and statistical physics have generating functions whose pole variety has a singularity with nontrivial monodromy. In the remainder of this section I will explain how hyperbolicity and the resulting deformations allow explicit asymptotic evaluation of some of these generating functions.

So as to keep the conversation more concrete, we consider as a running example the generating function for the probability of a Northgoing diamond in a uniform random tiling of the Aztec Diamond. This example is taken from [**BP11**, Section 4]. We have

$$\begin{aligned} F(X,Y,Z) &= \frac{Z/2}{(1-(X+X^{-1}+Y+Y^{-1})Z+Z^2)(1-YZ)} \\ &= \sum_{r,s,t} a_{r,s,t} X^r Y^s Z^t \end{aligned}$$

where a_{rst} is the probability that the domino covering the square (r,s) in the order t Aztec Diamond is oriented in a Northgoing direction. Figure 9 illustrates the Aztec Diamond shape of order 4 and the macroscopic features of a random tiling by dominoes of a larger Aztec Diamond (order 47).

Observe that r and s can be both positive and negative, while t is always positive and at least $|r|+|s|$. The denominator $Q(\mathbf{Z})$ is a Laurent polynomial whose zero set in $(\mathbb{C}^*)^d$ has two isolated singularities at $\pm(1,1,1)$. Near each of these, the zero sets of the two factors in the denominator look like a cone and a plane respectively: letting $Q = JH$, where J is the quadratic and $H = 1 - YZ$ is the log-linear term, we have

$$\mathsf{hom}(Q,(1,1,1)) = (Z^2 - \frac{1}{2}(U^2+V^2))(Y-Z)\,.$$

The zero set of this homogeneous polynomial is shown in Figure 10.

Let T' denote the flat torus $\mathbf{x} + \mathbb{R}^d/(2\pi\mathbb{Z})^d$ where $\mathbf{x}$ is any element of B, the component of $\mathsf{amoeba}(Q)^c$ over which the series converges. Changing variables via $\mathbf{Z} = \exp(\mathbf{z}) = (e^{z_1}, e^{z_2}, e^{z_3})$ and $d\mathbf{z} = d\mathbf{Z}/\mathbf{Z}$, then writing $\mathbf{z} = \mathbf{x} + i\mathbf{y}$ and $f(\cdot) = F \circ \exp(\mathbf{x} + i\cdot)$, yields

$$a_{\mathbf{r}} = (2\pi)^{-d} e^{-\mathbf{r}\cdot\mathbf{x}} \int_{T'} \exp(-i\mathbf{r}\cdot\mathbf{y}) f(\mathbf{y})\, d\mathbf{y}\,. \tag{3.4}$$

Up to this point the expression for $a_{\mathbf{r}}$ is exact. One can show that approximating f by $\mathsf{hom}(f)$ and T' by $\mathbb{R}^d$ does not change the leading asymptotic

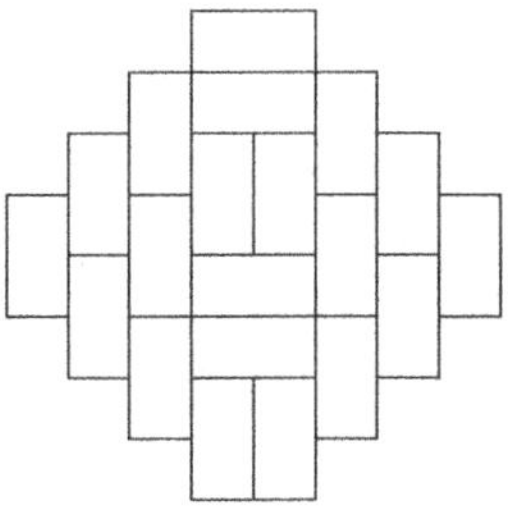

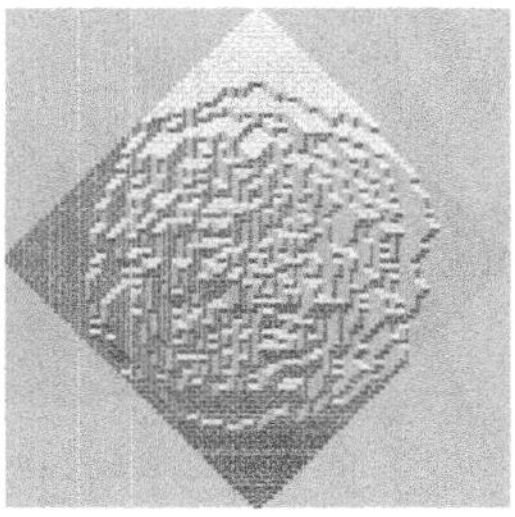

FIGURE 9. An Aztec diamond and a random tiling of another Aztec diamond.

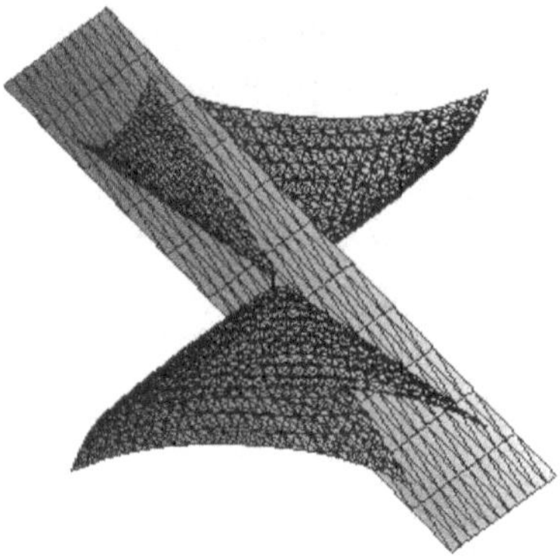

FIGURE 10. The homogenized zero variety near $(1, 1, 1)$.

term. This is a somewhat lengthy verification and it is here that the conical deformation of Corollary 3.10 is required. The upshot is that $a_{\mathbf{r}} = E(\mathbf{r})$ where E is the inverse Fourier transform of $\mathsf{hom}(f)$. For quadratics, the inverse Fourier transform is the dual quadratic. For a product of a quadratic and a log-linear function, an explicit computation yields [**BP11**, Theorem 3.9]

$$C \arctan \left(\frac{\sqrt{A^*(\mathbf{r}, \mathbf{r})}\sqrt{-A^*(\ell, \ell)}}{A^*(\mathbf{r}, \ell)} \right)$$

where A^* is the quadratic form dual to $\mathsf{hom}(f)$ and ℓ is the the log-linear factor viewed as an element of $(\mathbb{R}^d)^*$ in the logarithmic space. For example, when F is the Aztec generating function one obtains the following result, a pictorial version of which is given in Figure 11.

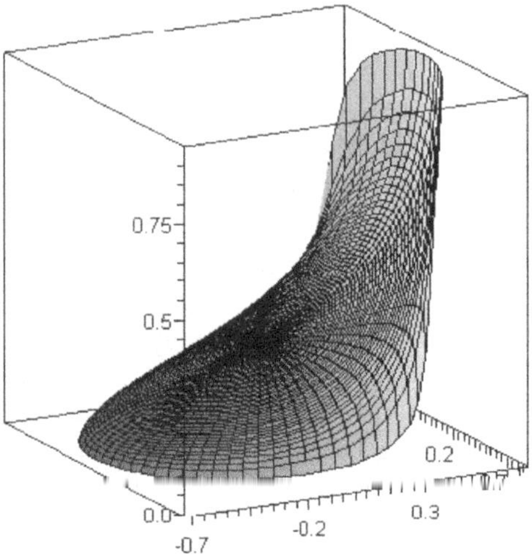

FIGURE 11. Scaling limit of $a_{r,s,t}$.

THEOREM 3.11 ([**BP11**, Theorem 3.7]). *The Northgoing placement probabilities* $\{a_{rst}\}$ *for the Aztec Diamond are asymptotically given by*

$$a_{rst} \sim \frac{1}{\pi} \arctan \left(\frac{\sqrt{t^2 - 2r^2 - 2s^2}}{t - 2s} \right)$$

when (r, s, t) *is in the cone* $\{(r, s, t) : t^2 > 2r^2 + 2s^2\}$ *and* $r + s + t$ *is odd.*

REMARK. The way the generating function is indexed, $a_{rst} = 0$ when $r + s + t$ is even. Asymptotics in the corners of the diamond outside the inscribed circle are given by $a_{rst} \to 1$ in one corner and $a_{rst} \to 0$ in the other three, where convergence occurs at an exponential rate as $t \to \infty$.

Part II: Stability

We now turn to the notion of stability. As we will see in Sections 4.3 and 5, the most useful properties of stability are closure properties: stability is preserved by a wealth of operations that are natural from an algebraic or probabilistic point of view. Stability is defined for general multivariate complex polynomials. Some of these properties or their proofs simplify when restricted to certain subclasses, such as polynomials whose coefficients are real or positive, or **multi-affine** polynomials, whose degrees in each variable never exceed 1. Even in these cases, however, some properties are proved only by going through the more general setting of complex coefficients.

4. Stability theory in one variable

As usual, the univariate theory is older and simpler. As with hyperbolicity, the origins of stability theory are in differential equations and control theory. After discussing this, we turn to combinatorial uses of the univariate theory. Stable generating polynomials produce coefficient sequences satisfying Newton's inequalities, implying, among other things, log-concavity. We end the section on univariate stability with a discussion of stability preserving operations and the so-called Laguerre–Pólya class.

4.1. Stability over general regions. In this section we will begin by thinking more generally of polynomials that avoid roots in some region Ω. We will call these Ω-stable. Throughout the remainder of this paper we will let

$$\mathcal{H} := \{z : \Im\{z\} > 0\} .$$

denote the open upper half-plane. I will always use the term "stable" to mean $\mathcal{H}$-stable, as in Definition 1.2, and will use "Hurwitz stable" or "Ω-stable" for other notions of stability. Recalling G.-C. Rota's philosophical observation, one might keep in mind that the Riemann Hypothesis is equivalent to Ω-stability where $\Omega = \{z : \Re\{z\} \neq 1/2\}$. This may seem like a strained connection, but in fact some of the literature we review in Section 7 is explicitly motivated by the desire to understand the zeta function.

Statistical physicists have a different motive for understanding regions free of zeros. To explain this, we examine some graph theoretic models, paraphrasing the exposition in [**Sok01**]. Let $G = (V, E)$ be a finite graph and let q be a positive integer. Define a polynomial in variables $\{x_e : e \in E\}$

$$Z(q, \{x_e : e \in E\}) := \sum_{\sigma} \prod_{e=\overline{vw}\in E} \left(1 + x_e \delta_{\sigma(v),\sigma(w)}\right)$$

where σ ranges over q-colorings of the vertices, that is, all maps from V to $\{1, \ldots, q\}$. This is elementarily seen to be equivalent to the alternative definition

$$Z(q, \{x_e : e \in E\}) := \sum_{E' \subseteq E} q^{k(E')} \prod_{e \in E'} x_e$$

where $k(E')$ denotes the number of connected components in the subgraph (V, E'). The last formulation makes it obvious that Z is a polynomial in q as well as in $\{x_e\}$. When $x_e = -1$ for all e this specializes to the chromatic polynomial $\chi(q)$; more generally, taking $x_e = x$ for all e gives the bivariate **Tutte polynomial**.

The polynomial Z is also the **partition function** for the q-state Potts model. Statistical physicists are interested in **phase transitions** where the behavior of the system depends non-analytically on its parameters. Being a polynomial, Z is of course analytic in all its parameters. Typically though, one is interested in infinite-volume limits such as the free energy

$$f(q, x) := \lim_{n \to \infty} n^{-1} \log Z_n(q, x)$$

where Z_n is taken on a graph G_n in a family of graphs whose increasing limit is $\mathbb{Z}^d$. The function f may fail to be analytic at (q, x) if the zeros of $Z_n(q, x)$ have a limit point at (q, x). Therefore, there is physical significance in keeping the zeros of the polynomial Z_n out of specified regions. Such results are often known as **Lee-Yang** theorems, after [**YL52, LY52**] (see also [**LS81**]).

One such theorem was proved by D. Wagner in 2000. The **reliability polynomial** R_G of the graph $G = (V, E)$ is the probability that the graph is connected when each edge is kept or deleted independently with probability x, which is a polynomial in the parameter x. Brown and Colbourn conjectured [**BC92**] that all zeros of $R_G(x)$ are in the closed disk $\{z : |z-1| \leq 1\}$. In other words, R_G is $\mathcal{D}^c$-stable where $\mathcal{D} := \{z : |z-1| \leq 1\}$. This was proved for the class of series-parallel graphs in [**Wag00**, Theorem 0.2]. A simpler proof was found by Sokal [**Sok01**, Section 4.1, Remark 3], who actually proved the stronger multivariate stability result. On the other hand, Royle and Sokal [**RS01**] showed that the original univariate conjecture is false for general graphs; they also showed that multivariate $\mathcal{D}^c$-stability of R_G holds *if and only if* G is series-parallel. Sokal [**Sok11**] has recently conjectured that R_G is (univariate) $\mathcal{D}^c$-stable for the complete graphs, K_n.

Hurwitz stability. The association of zero-free regions with the term "stability" originated in ODE's and control theory. Its use is attributed to Hurwitz. In fact, Ω-stability with $\Omega = \{z : \Re\{z\} \geq 0\}$ is sometimes called "Hurwitz stability". In ODE's, it is easy to see the physical significance of Hurwitz stability. Let M be a matrix and consider the linear system $\mathbf{y}' = M\mathbf{y}$. Hurwitz stability is equivalent to all eigenvalues of M having negative real parts, which is equivalent to all homogeneous solutions decaying, and hence good control over the system $\mathbf{y}' - M\mathbf{y} = \mathbf{v}(t)$. One may also consider discrete-time analogues, such as the system $\mathbf{y}^{(n+1)} - Q\mathbf{y}^{(n)} = \mathbf{v}^{(n)}$ where Q would correspond to the exponential of M in the previous system. Now decay of homogeneous solutions is equivalent to Ω-stability when Ω is the complement of the open unit disk; polynomials whose zeros are all in the open unit disk are said to be **Schur-stable**.

For a less trivial example we turn to control theory. Following [**Hen91**, Section 10.3], we consider systems which turn an input signal $f(t)$ into an output $u(t)$. Many systems, and in particular those built from networks of impedances, are not only linear but also time homogeneous, that is, the map $f \mapsto u$ commutes with time translation. In terms of Laplace transforms, this means that the system acts multiplicatively, meaning that if $\mathcal{L}$ denotes the Laplace transform, then such systems obey the law

$$\mathcal{L}u = g \cdot \mathcal{L}f \tag{4.1}$$

for some function $g : [0, \infty) \to \mathbb{R}$ that is called the **transfer function**. For example an L-C-R circuit, with an input voltage function $f(t)$ between the inductor and the resistor producing an output $u(t)$ of current from the inductor to the capacitor, satisfies (4.1) with $g(s) = (R + Ls + C^{-1}/s)^{-1}$. In impedance networks the transfer function is in fact always rational, $g(s) = p(s)/q(s)$ for some polynomials p and q. Generally, a system is said to be stable if bounded input produces bounded output.

To check whether the system with transfer function g is stable, it suffices to test it on inputs of the form $f(t) = e^{i\omega t}$, the Laplace transform of which is $(\mathcal{L}f)(s) = 1/(s - i\omega)$. Let the poles of the rational function g be $a_1, \ldots, a_r$. The $\mathcal{L}u = g \cdot \mathcal{L}f$ has poles $a_1, \ldots, a_r, i\omega$. We may write $\mathcal{L}u$ in a partial fraction expansion resulting in $\sum_{j=0}^r p_j/(s - a_j)$ for some polynomials $\{p_j\}$; here we have set $a_0 := i\omega$. Inverting the Laplace transform gives a sum $\sum_{j=1}^r P_j(t)e^{a_j t}$ for some collection $\{P_j\}$ of polynomials. This is bounded if and only if each a_j is either in the open left half-plane or is imaginary and a simple zero. But when g has an imaginary zero a_j, then setting $a_0 = a_j$ (that is, taking the input to be $e^{i a_j t}$) produces a doubled root and an unbounded output. Hence, bounded inputs produce bounded outputs if and only if all poles of g lie strictly in the left half-plane.

We conclude that stable behavior of the system corresponds to Hurwitz stability of the denominator of the transfer function g. For example, in the L-C-R circuit above, the denominator of g is a quadratic with positive coefficients. The real parts of the roots are always negative, whence such a system always behaves stably.

Differential equations of order d may be transformed into first order systems in d variables by the well known trick of representing the first $d - 1$ derivatives as new variables. It is not surprising, therefore, that Hurwitz stability also arises in the stability analysis of order-d linear differential equation with constant coefficients. Given d initial conditions and an inhomogeneous term, we may write such an equation as

$$f^{(d)} + a_{d-1} f^{(d-1)} + \cdots + a_0 f = H \tag{4.2}$$

with initial conditions $f^{(j)}(0) = b_j$ for $0 \leq j \leq d-1$. Assuming H to grow at most exponentially, a Laplace transform exists near the origin. We may take the Laplace transfrom of both sides of (4.2). Using linearity and the rule $\mathcal{L}f'(s) = s\mathcal{L}f(s) - f(0^+)$ and denoting $g := \mathcal{L}f$ gives inductively

$$\mathcal{L}(f^{(d)})(s) = s^d g(s) - s^{d-1} f(0^+) - s^{d-2} f'(0^+) - \cdots - f^{(d-1)}(0^+) .$$

Plugging this and the boundary conditions into (4.2) yields the following equation for g:

$$\begin{aligned} s^d g(s) - s^{d-1} b_0 - s^{d-2} b_1 - \cdots - b_{d-1} & \\ + a_{d-1}\left[s^{d-1} g(s) - s^{d-2} b_0 - \cdots - b_{d-2} \right] & \\ \vdots \quad & \\ + a_1 \left[s^1 g(s) - b_0 \right] & \\ + a_0 g(s) & = \mathcal{L}H(s) . \end{aligned}$$

Letting $p(s) := s^d + a_{d-1} s^{d-1} + \cdots + a_0$ denote the characteristic polynomial of the equation (4.2), and $p_k(s)$ denote the shifted polynomial $s^k + a_{d-1} s^{k-1} + \cdots + a_{d-k}$, we may rewrite the equation for g as

$$p \cdot g = \mathcal{L}H + \sum_{j=0}^{d-1} b_j p_{d-1-j}$$

and hence

$$g(s) = \frac{\mathcal{L}H(s) + \sum_{j=0}^{d-1} b_j p_{d-1-j}(s)}{p(s)} . \tag{4.3}$$

Again, poles of g with positive real part produce unbounded output, as do purely imaginary poles $i\omega$ when the driving term is taken to be $e^{i\omega t}$, whose Laplace transform $\mathcal{L}H$ has a pole at $i\omega$. Thus, again, Hurwitz stability is equivalent to bounded output on bounded input.

4.2. Real roots and Newton's inequalities. I will now return, permanently, to upper half-plane stability, which will be called, simply, "stability". In one variable, when the coefficients of f are real, the zeros come in conjugate pairs. Stability, therefore, is equivalent to having only real zeros. When the coefficients are nonnegative, a strictly positive zero is impossible, whence stability is further equivalent to having all zeros on the negative half line.

Combinatorialists have long sought to exploit the properties of stable generating functions. The universal problem in combinatorial enumeration is to count a family of structures indexed by one or more positive integer parameters. In the case of one parameter, say n, the sequence $\{a_n\}$ of counts and the corresponding generating function $f(z) := \sum_n a_n z^n$ are of fundamental interest. In the setting of probability generating functions, the coefficients a_n are nonnegative and sum to one. In this case, if f has only real roots, then the distribution which gives probability a_n to the value n is representable as the sum of independent random variables each taking the value 0 or 1 (Bernoulli random variables). There may or may not be a natural interpretation for these Bernoulli variables; see [**HKPV09**, Example 23] for an example in which there can be no natural interpretation.

Often it is intuitively plausible that the sequence $\{a_n\}$ is **unimodal**, meaning that for some k, we have $a_0 \leq \cdots \leq a_k \geq a_{k+1} \geq a_{k+2} \geq \cdots$. The enumeration literature is littered with examples in which unimodality is conjectured (see for example the survey [**Sta89**] and the follow-up to this [**Bre94**]), but a proof is often elusive for the reason that there is no obvious theoretical framework within which to prove unimodality. There are, however some stronger properties for which natural avenues of proof exist.

DEFINITION 4.1 (log-concavity). *A finite or infinite sequence $\{a_k\}$ of nonnegative numbers is said to have no internal zeros if the indices of the nonzero terms form an interval $[r, s]$. The sequence is said to be* ***log-concave*** *if it has no internal zeros and if*

$$a_{k-1}a_{k+1} \leq a_k^2$$

for $k \in [r+1, s-1]$. The sequence $\{a_r, \ldots, a_{r+k}\}$ with no internal zeros is said to be ***ultra log-concave*** *if*

$$\left(\frac{a_j}{\binom{k}{j}}\right)^2 \geq \frac{a_{j+1}}{\binom{k}{j+1}} \frac{a_{j-1}}{\binom{k}{j-1}} .$$

It is immediate that ultra log-concavity implies log-concavity which implies unimodality. While ultra log-concavity appears to be the least natural of these properties, it was shown three centuries ago by Newton to follow from stability of the generating function.

THEOREM 4.2 (Newton's inequalities). *Suppose $f(z) := \sum_{k=0}^n a_k z^k$ is a real stable polynomial. Then for $1 \leq k \leq n-1$,*

$$\left(\frac{a_k}{\binom{n}{k}}\right)^2 \geq \frac{a_{k+1}}{\binom{n}{k+1}} \frac{a_{k-1}}{\binom{n}{k-1}} . \tag{4.4}$$

If furthermore the coefficients a_k are all nonnegative then the roots of f are all nonpositive and the sequence is ultra log-concave.

PROOF: By Rolle's Theorem, if a univarite polynomial f has only real zeros then so does its derivative. This observation will be useful many times below.

We use it now to deduce that

$$Q(z) := \left(\frac{d}{dz}\right)^{k-1} f(z)$$

has only real zeros. Reversing a sequence of coefficients via $R(z) := z^{n-k+1} Q(1/z)$ also preserves the property of having all real roots. By Rolle's Theorem again, $S(z) := (d/dz)^{n-k+1} R(z)$ has all real roots. But $S(z)$ is the trinomial

$$\frac{n!}{2}\left(\frac{a_{k-1}}{\binom{n}{k-1}} z^2 + 2\frac{a_k}{\binom{n}{k}} z + \frac{a_{k+1}}{\binom{n}{k+1}}\right),$$

and the theorem follows from the discriminant test for quadratics. □

A curious result in a similar vein was proved by Gurvits [**Gur08**, Lemma 3.2]; the proof, which is a few lines of calculus, is omitted.

PROPOSITION 4.3. *Let f be a probability generating polynomial of degree d and let $C := \inf f(t)/t$. If f is stable then*

$$a_1 = f'(0) \geq \left(\frac{d-1}{d}\right)^{d-1} C.$$

Equality holds if and only if $f = (1-q+qt)^d$ generates a binomial distribution. The value $[(d-1)/d]^{d-1}$ increases to e^{-1} as $d \to \infty$; for infinite series, the bound $a_1 \geq e^{-1}C$ holds with equality if and only if $f(t) = e^{\lambda(t-1)}$ is the generating function for a Poisson distribution. □

There are a number of techniques that can be used to establish stability (also known, in the univariate case, as the real root property). In some cases one has a sequence $\{f_n\}$ where f_n has degree n and the roots of f_{n-1} interlace the roots of f_n, meaning that each interval between roots of f_n contains a root of f_{n-1}. If this is true, then often it is possible to prove it by induction. The following example from [**Sta89**] illustrates this.

EXAMPLE 4.4 (Hermite polynomials). *The Hermite polynomials are a sequence of polynomials defined by*

$$H_n(z) := \sum_{k=0}^{\lfloor n/2 \rfloor} \frac{(-1)^k n!(2z)^{n-2k}}{k!(n-2k)!}.$$

The are orthogonal with respect to the Gaussian measure $e^{-x^2}\,dx$. They satisfy the recursion

$$H_n(z) = -e^{z^2} \frac{d}{dz}\left(e^{-z^2} H_{n-1}(z)\right).$$

Assume for induction that $H_{n-1}(z)$ has $n-1$ real zeros. It is clear from this that H_n has $n-2$ zeros interlacing the zeros of H_{n-1}. To see that H_n also has a zero less than all the zeros of H_{n-1}, observe that $e^{-z^2} H_{n-1}$ tends to *zero as $z \to -\infty$, therefore $e^{-z^2} H_{n-1}$ has an extreme value to the left of its leftmost zero, which is a zero of H_n. Similarly, H_n has a zero greater than the greatest zero of H_{n-1}, and the induction is established.*

There is a more methodical way that often works to prove interlacing by induction. If f and g are real polynomials of respective degrees n and $n-1$ having all real roots and the zeros of g interlace the zeros of f, it may be seen elementarily that $f + \lambda g$ has n real zeros for any real λ. A simple application of this idea is to the sequence $\{g_n\}$ defined by the three-term recurrence

$$g_{n+1}(x) = axg_n(x) + bg_{n-1}(x) ,$$

which include the Chebyshev polynomials and Laguerre polynomials (a and b depending on n in the latter case). The same idea is the basis for a theorem independently proved by Heilmann and Lieb [**HL72**], by Gruber and Kunz [**GK71**] and, in part, by Nijenhuis [**Nij76**]. A matching on a finite weighted graph or multi-graph is a subset M of the edge set such that any two edges in M are disjoint (share no vertex). Fix a graph $G = (V, E)$ and let $\{W(e) : e \in E\}$ be a set of nonnegative weights. Let a_j denote the number of matchings with j edges, counted by weight, where the weight of a matching is the product of the weights of the edges in the matching; by convention we take $a_0 : -1$. In the simplest case, $W(e) \equiv 1$ and a_j simply counts matchings of size j, size being the number of edges.

If we are allowed to set $W(e) = 0$ for some edges, then the complete graph is universal. We therefore assume that $G = K_n$ for some n.

THEOREM 4.5 (matchings). *The number of weighted matchings of K_n enumerated by size is ultra log-concave.*

PROOF: We show stability of a certain generating function, the most convenient being

$$Q_n(z) := \sum_{j=0}^{\lfloor n/2 \rfloor} (-1)^j a_j z^{n-2j} . \tag{4.5}$$

This is monic, of degree n, and is odd or even depending on n. If we show Q_n is stable and has n distinct roots, then letting $Q_n(z) = R_n(z^2)$ in the even case and $Q_n(z) = zR_n(z^2)$ in the odd case, it follows that $R_n(-z)$ has $\lfloor n/2 \rfloor$ negative real roots, hence is stable; it is also the generating function for matchings by size, hence the proof will be complete.

Taking limits at the end, we may assume without loss of generality that the weights $W(e)$ are all strictly positive. To see that Q_n is stable, let Q_n^S denote the generating function defined by (4.5) on the graph $K_n \setminus S$. We use the recursion

$$Q_n(z) = zQ_n^u(z) - \sum_{v \neq u} W(u, v) Q_n^{u,v}(z) \tag{4.6}$$

which is obvious from the definition. The induction hypothesis is that Q_n^u is stable with distinct roots, which are interlaced by the roots of $Q_n^{u,v}$ for any $v \neq u$.

To verify the induction, we consider the sign of Q_n at the $n-1$ zeros of Q_n^u in decreasing order. By the interlacing property, the sign of each $Q_n^{u,v}$ alternates, starting out positive because $Q_n^{u,v}(+\infty) = +\infty$. Therefore, $zQ_n^u - \sum_v W(u,v)Q_n^{u,v}$ alternates, starting out negative. This implies the existence of $n-2$ roots of Q_n interlaced by the roots of Q_n^u. But also there is one root of Q_n to the right of every root of Q_n^u because Q_n is negative at the rightmost root of Q_n^u and positive at $+\infty$. Similarly there is one root of Q_n to the left of every root of Q_n^u. This completes the inductive proof. □

A second method by which one can establish univariate stability is via closure properties of the class of univariate stable polynomials. This topic will be expanded in the next section, but for now we mention a result from [**Bre89**], whose proof we will omit. Let $(x)_i$ denote the falling product $x(x-1)\cdots(x-i+1)$ and let $(x)^i$ denote the rising product $x(x+1)\cdots(x+i-1)$.

THEOREM 4.6 ([**Bre89**, Theorems 2.4.2–2.4.3]). *Let $f(x) := \sum_{k=0}^n a_k x^k$ be real stable with nonnegative coefficients. Then the polynomials $\sum_{k=0}^n a_k (x)_k$ and $\sum_{k=0}^n a_k (x)^k$ are real stable as well.* □

In the following example, a function from the set $\{1, \ldots, n\}$ to itself is represented as a directed graph with an edge from j to $f(j)$ for each j.

EXAMPLE 4.7 (functions enumerated by components). *Let $b(n,k)$ be the number of functions from the set $\{1, \ldots, n\}$ to itself whose directed graph has precisely k components. It is elementary, cf.* [**GJ83**, Example 3.3.28], *that*

$$b(n,k) = \sum_{i=1}^n \binom{n-1}{i-1} n^{n-i} c(i,k)$$

where $c(i,k)$ is a signless Stirling number of the first kind. Multiplying by x^k and summing over k gives

$$f_n(x) := \sum_{k=1}^n b(n,k) x^k = \sum_{i=1}^n \binom{n-1}{i-1} n^{n-i} (x)^i .$$

The corresponding sum replacing $(x)^i$ by x^i has the short closed form

$$\sum_{i=1}^n \binom{n-1}{i-1} n^{n-i} x^i = x(x+n)^{n-1} .$$

Evidently this has only real zeros, so applying the conclusion of Theorem 4.6, we see that f_n is stable as well.

A third method is an equivalent characterization for ultra log-concave that is due to Edrei [**Edr53**]. This relies on the notion of **total positivity**, a theme developed at length by Karlin [**Kar68**] and for which a number of combinatorial applications are given in [**Bre95**].

DEFINITION 4.8 (Total positivity; Pólya frequency sequence).
(i) An infinite matrix is said to be totally positive (TP) if all of its minors are nonnegative.
(ii) A sequence $\{a_n\}$ is said to be a Pólya frequency sequence (PF-sequence) if the matrix $A_{nk} := a_{n-k}$ is totally positive. Here we take $a_i := 0$ if $i < 0$ or the sequence $\{a_n\}$ is finite and has length less than i.

A proof of the following equivalence may be found in in [**Kar68**, Theorem 5.3].

THEOREM 4.9 (Edrei's equivalence theorem). *The nonnegative sequence $(a_0, \ldots, a_d)$ is a PF-sequence if and only if the generating polynomial $\sum_{k=0}^d a_k z^k$ is stable.* □

Brenti [**Bre94**, Section 3] points out that only the 2×2 minors are required for log-concavity, hence unimodality. Nevertheless, nonnegativity of all minors has a combinatorial interpretation which is given in [**Bre95**, Theorem 3.5]. This interpretation is somewhat abstract, but in special cases the interpretation can be more concrete.

EXAMPLE 4.10 (r-derangements). *A permutation of $\{1, \ldots, n\}$ is said to be an r-derangement if all its cycles have length at least r. Thus a 1-derangement is any permutation, a 2-derangement is a classical derangement, and so forth. Let $b(n,r)$ count the number of r-derangements of $\{1, \ldots, n\}$. Brenti* [**Bre95**, Corollary 5.9] *shows that $f_n := \sum_{r=1}^n b(n,r) z^r$ is stable. To do so, he shows that a related sequence $\{c_r(n)\}$ is a PF-sequence, hence stable, then applies Theorem 4.6 to recover the result with $b(n,r)$ in place of $c_r(n)$.*

We end this section with one more example of a stable generating polynomial, this one taken from [**Sta89**]. Recall from Example 2.4 that if A is a real symmetric matrix and B is real positive semi-definite then $f(z) := \mathrm{Det}\,(A + zB)$ is a real stable polynomial.

EXAMPLE 4.11 (spanning forests enumerated by component). *Let G be a finite graph with possibly multiple edges and let*

$$a_k(G) := \sum_F \gamma(F)$$

where the sum is over all spanning forests F of G such that F has precisely k components, where $\gamma(F)$ is the product of the cardinalities (numer of vertices) of the k components. The factor of γ changes the eumeration from forests to rooted forests. Following [**Sta89**, Proposition 4], *let us see that the polynomial $f_G := \sum_k a_k(G) z^k$ is stable. Let A be the matrix whose rows and columns are indexed by the vertices of G with entry $A_{ij} := \deg(u)$ if $u = v$ and otherwise equal to $-N(i,j)$ where $N(i,j)$ is the number of* *edges between i and j. It is known that $f_G = \mathrm{Det}\,(A + zI)$. It follows that f_G is stable. Therefore, the number of rooted forests enumerated by components is ultra log-concave, hence unimodal. Stanley attributes the identity $f_G = \mathrm{Det}\,(A + zI)$ to Kelmans.*

4.3. The Laguerre–Pólya class. Suppose we transform a polynomial $f(z) = \sum_{k=1}^{n} a_k z^k$ by multiplying each coefficient a_k by a specified constant λ_k. We may ask which sequences $\boldsymbol{\lambda} := \{\lambda_k\}$ are **multiplier sequences**, meaning that the resulting operator $T_{\boldsymbol{\lambda}}$ preserves the class of real stable polynomials. A classical theorem due to Pólya and Schur [**PS14**] gives a complete and beautiful characterization of multiplier sequences.

THEOREM 4.12 (Pólya-Schur 1914). *Let $\boldsymbol{\lambda} = \{\lambda_n : n \geq 0\}$ be a sequence of real numbers and let $T_{\boldsymbol{\lambda}}$ denote the linear operator defined by*

$$T_{\boldsymbol{\lambda}}\left(\sum_{k=0}^{n} a_k z^k\right) = \sum_{k=0}^{n} \lambda_k a_k z^k .$$

Denote by Φ the formal power series

$$\Phi(z) := \sum_{k=0}^{\infty} \frac{\lambda_k}{k!} z^k .$$

Then the following are equivalent.

(i) $\boldsymbol{\lambda}$ is a multiplier sequence;
(ii) Φ is an entire function and is the limit, uniformly on compact sets, of the polynomials with all zeros real and of the same sign;
(iii) Φ is entire and either $\Phi(z)$ or $\Phi(-z)$ has a representation

$$C z^n e^{\alpha_0 z} \prod_{k=1}^{\infty} (1 + \alpha_k z)$$

where n is a nonnegative integer, C is real, and α_k are real, nonnegative and summable.
(iv) For all nonnegative integers n, the polynomial $T_{\boldsymbol{\lambda}}[(1+z)^n]$ is real stable with all roots of the same sign.

We will not prove the general Pólya-Schur theorem here, proving only some special cases later as we need them. Some examples and remarks will clarify its meaning and possible uses. Condition (iv) may be thought of as saying that the polynomials $(1 + z)^n$ are universal test cases for stability preserving: a multiplier sequences preserving stability of these will preserve stability of all real stable polynomials.

EXAMPLE 4.13 (dilation). *If b is a nonnegative integer, setting $\lambda_k := b^k$ produces the operator $T_{\boldsymbol{\lambda}} f(z) = f(bz)$. Clearly this preserves stability. The representation in (iii) is obvious because $\Phi(z) = e^{bz}$.*

EXAMPLE 4.14 (factorials). *If $n \geq 1$ is an integer then the sequence $\lambda_k := (n)_k := n!/(n-k)!$, defined to vanish when $k > n$, produces the exponential generating function $\Phi(z) = (1+z)^n$. By criterion (iii) this is a multiplier sequence. Dividing by the constant $n!$ we see that $\lambda_k = 1/(n-k)!$ for $k \leq n$ and zero for $k > n$ defines a multiplier sequence. For real polynomials of degree n, stability of f is equivalent to stability of the inversion $z^n f(1/z)$, hence $\lambda_k = 1/k!$ defines a multiplier sequence on polynomials of degree n. This is true for every n, whence $\{1/k!\}_{k=0}^{\infty}$ is a multiplier sequence. This result is due to Laguerre.*

EXAMPLE 4.15 (coefficientwise multiplication). *Applying the previous example to a polynomial $g(z) = \sum_{k=0}^{n} a_k z^k$ with negative real roots shows that $\sum_{k=0}^{n} (a_k/k!) z^k$ also has negative real roots. Setting $\lambda_k := a_k$ for $k \leq n$ and zero for $k > n$ we see that its exponential generating function $\Phi(z) = \sum_{k=0}^{n} a_k z_k / k!$ is of the form in (iii) with $\alpha_0 = 0$. We conclude that $\{a_k\}$ is a multiplier sequence. In other words, if we multiply term by term the coefficient sequences of two real rooted polynomials, at least one of which has nonnegative coefficients, we get another real rooted coefficient sequence. Thus the real root property for polynomials with nonnegative coefficients is closed under* ***Hadamard products****.*

Multiplier sequences are related to the so-called **Laguerre–Pólya** class. An entire function is said to be in the Laguerre–Pólya class, denoted $\mathcal{L}\text{–}\mathcal{P}$, if it is the limit, uniformly on compact subsets of $\mathbb{C}$, of polynomials with only real zeros; the terminology goes back at least to [**Sch47**]. It is said to belong to the subclass $\mathcal{L}\text{–}\mathcal{P}I$ if it is the limit of polynomials with only real zeros all of the same sign. Thus Theorem 4.12 asserts that generating functions of multiplier sequences are precisely the functions of class $\mathcal{L}\text{–}\mathcal{P}I$.

A related notion to that of a multiplier sequence is the notion of a **complex zero decreasing sequence** (CZDS). Let $Z_c(f)$ denote the number of non-real zeros of f. Say that a finite or infinite sequence $\{a_k\}$ is a CZDS if for any real polynomial $f(z) = \sum_{k=0}^{n} b_k z^k$,

$$Z_c\left(\sum_{k=0}^{n} a_k b_k z^k\right) \leq Z_c(f)\,. \tag{4.7}$$

In particular, in order for $\{a_k\}$ to be a CZDS, a value of zero on the right of (4.7) (no non-real zeros) implies no non-real zeros on the left, so any CZDS is a multiplier sequence. The converse, however, is not true.

To see that there is any nontrivial CZDS, we observe that the operator $z(d/dz)$ is represented by the sequence $a_k := k$ and can never increase the number of non-real zeros. Although not every multiplier sequences is a CZDS, each multiplier sequence leads to a CZDS via the following result going back to Laguerre.

PROPOSITION 4.16. *Let $\Phi \in \mathcal{L}\text{–}\mathcal{P}$ have zeros only in $(-\infty, 0]$. Then the sequence $\{\Phi(k) : k \geq 0\}$ is a CZDS.* □

Related to the notions of multiplier sequences and CZDS is the notion of a multiplier sequence for the property of being nonnegative on real inputs. Say that $\{c_k\}$ is a Λ-sequence if multiplication of coefficients term by term preserves the property of being everywhere nonnegative:

$$\sum_{k=0}^{n} a_k x^k > 0 \text{ for all } x \in \mathbb{R} \Longrightarrow \sum_{k=0}^{n} c_k a_k x^k > 0 \text{ for all } x \in \mathbb{R}\,.$$

Note that negative values are allowed for the numbers c_k. The following classical relationship is proved in [**Wid41**].

PROPOSITION 4.17. *If $\{a_k\}$ is a CZDS then $\{a_k^{-1}\}$ is a Λ-sequence.* □

Connections abound between these notions: stability, multiplier sequences, the Laguerre–Pólya class, Λ-sequences, etc., which we will not have time to survey here; the reader is referred to [**CC95**] for an introduction. Λ-sequences, for example, are neatly characterized by a determinant condition and also by the so-called Hamburger moment problem: $\{c_n\}$ is a Λ-sequence if and only if $c_n = \int_{-\infty}^{\infty} t^n \, d\mu(t)$ for some measure μ on $\mathbb{R}$ not supported on finitely many points. At the root of much of the present interest about these properties is their connection to the Riemann hypothesis. An account is given in [**CV90, Cso03**]. Finally, a small related literature on nonlinear transformations that preserve real stable polynomials may be of interest. It was conjectured independently by S. Fisk and R. Stanley that if the real polynomial $\sum_k a_k z^k$ has all real roots then so does the polynomial whose z^k coefficient is $a_k^2 - a_{k-1}a_{k+1}$. Some progress was made by McNamara and Sagan in [**MS10**]; a proof of the conjecture and more was recently found by Brändén [**Brä11**].

5. Multivariate stability

Let $\mathcal{H}$ denote the open upper half-plane; thus stability is equivalent to having no zeros in the region $\mathcal{H}^d$. We remark on some connections to other notions of stability. For homogeneous polynomials, the zero set is invariant under multiplication by $e^{i\theta}$ in each coordinate, therefore all notions of half-plane stability coincide. Any **circular** region (the interior or exterior of a circle) is transformed into $\mathcal{H}$ by a Möbius transformation, which allows the theory of Ω-stable functions to be mapped to the ordinary theory of stable functions via a bi-rational change of variables whenever Ω is the product of circular regions. This mapping acts nicely with respect to some aspects such as closure properties but not as nicely with respect to coefficient sequences. Thus, for example, it may shown that **Schur-stability**, where $\Omega = \mathcal{D}^d$ and $\mathcal{D}$ is the open unit disk, has an unexpected closure property: if f and g are multi-affine Schur-stable polynomials then the **Hadamard product** $f \bullet g$ is Schur-stable as well; here the Hadamard product of $\sum_S a(S)\mathbf{z}^S$ and

$\sum_S b(S)\mathbf{z}^S$ is defined to be $\sum_S a(S)b(S)\mathbf{z}^S$. Our principal interest is in $\mathcal{H}$-stable polynomials and their coefficients. We will not henceforth consider notions of stability other than upper half-plane stability.

A number of the closure properties of the class of stable polynomials extend immediately from the univariate to the multivariate setting or are otherwise elementary. In particular, if $f \in \mathbb{C}[z_1, \ldots, z_d]$ is stable then so are the polynomials obtained by the following operations:

(a) dilation: replacing f by $f(b_1 z_1, \ldots, b_d z_d)$, where $\{b_j\}$ are nonnegative constants;
(b) permuting the names of the variables;
(c) specialization: setting z_j equal to a constant in $\mathcal{H}$;
(d) diagonalization: setting z_j equal to z_i;
(e) inversion: replacing f by $f(-1/z_1, z_2, \ldots, z_d)$;
(f) differentiation: replacing f by $\partial f / \partial z_j$;
(g) limits: $f_n \to f$ uniformly on compact sets and f_n stable implies f is stable;

see for example [**Wag11**, Lemma 2.4]; here, and after, in results such as the this one, we will take "stable" to include the zero polynomial in order not to have to make exceptions.

We also recall that Proposition 1.3 equates stability of a real homogeneous polynomial to hyperbolicity in all directions in the positive orthant. Although the definition of hyperbolicity is more complicated for non-homogeneous polynomials, the corresponding fact characterization of real stable polynomials is not.

PROPOSITION 5.1. *A (non identically zero) real d-variable polynomial f is stable if and only if for all $\mathbf{v} \in \mathbb{R}^d$ and $\mathbf{u} \in \mathbb{R}^d_+$, the polynomial $t \mapsto f(\mathbf{v} + t\mathbf{u})$ has only real zeros.*

PROOF: If f has a zero $(z_1, \ldots, z_d)$ in $\mathcal{H}^d$, then writing $z_j = v_j + iu_j$, we see that $\mathbf{u} \in \mathbb{R}^d_+$ and i is a zero of $f(\mathbf{v} + t\mathbf{u})$. Conversely, if $\mathbf{u} \in \mathbb{R}^d_+$ and $\mathbf{v} \in \mathbb{R}^d$ such that $f(\mathbf{v} + t\mathbf{u})$ has a root with nonvanishing imaginary part, we may conjugate if necessary to find a root $t = a + bi$ with $b > 0$; then $\mathbf{v} + t\mathbf{u}$ is a zero of f in $\mathcal{H}^d$. □

There is one geometric feature of functions of two or more complex variables, absent from the univariate theory, that is worth noting because it strengthens closure property (c). Let f be any function in d variables, vanishing at $\mathbf{b} = (b_1, \ldots, b_d)$ and analytic in a neighborhood of $\mathbf{b}$. Suppose that $f(\cdot, b_2, \ldots, b_d)$ is not the zero function. Then there is a continuous multi-valued function ϕ on a neighborhood U of $(b_2, \ldots, b_d)$ in $\mathbb{C}^{d-1}$ such that $\phi(\mathbf{z}) \to b_1$ as $\mathbf{z} \to (b_2, \ldots, b_d)$ and such that $f(\phi(\mathbf{z}), \mathbf{z}) = 0$. This perturbation property follows from the fact that at least one multivariate Puiseux series solution exists for z_1 as a function of $z_2, \ldots, z_d$ which converges to b_1. This leads to the following lemma.

LEMMA 5.2. *Let $B_j, 1 \le j \le d$ be open subsets of $\mathbb{C}$. Suppose a polynomial f is nonvanishing on $\prod_{j=1}^d B_j$. Then f is nonvanishing on $\prod_{j=1}^d \overline{B_j}$ except possibly in two cases: f can vanish on the product of boundaries $\prod_{j=1}^d \partial B_j$, and f can vanish at a point $(b_1, \dots, b_d)$ with coordinates $\{b_j : j \in S\}$ in the interior of B_j for some set $S \subseteq \{1, \dots, d\}$ provided that f vanishes identically for all values of those coordinates, that is, $f(b'_1, \dots, b'_d) = 0$ whenever $b'_j = b_j$ for all $j \notin S$.*

In particular, taking $B_j = \mathcal{H}$ for all j, we see that stability implies no zeros in $\overline{\mathcal{H}}^d$ except for real zeros or degenerate cases.

A further consequence of this lemma is that (c) can be strengthened to include setting z_j equal to a real constant.

PROOF: Let f vanish at $\mathbf{b} = (b_1, \dots, b_d)$ where $b_j \in \partial B_j$ for $j \le r$ and $b_j \in B_j$ for $j > r$. We need to show that the hypotheses are contradicted if $k > r$ is a coordinate such that f does not vanish identically when the remaining coordinates are fixed. This follows if we perturb each coordinates $b_j, j \ne k$ so as to lie in the open region B_j, by sufficiently small amounts so that the perturbation solution for z_k lies in B_k. □

The above facts are generally quite elementary. Before we get to the fun stuff, there is some serious overhead in deriving further closure properties. One route to developing properties of multivariate stable functions is via the notion of **proper position**. This is in some sense a multivariate version of interlacing of roots. The development here specifically avoids this because I do not find it intuitive and because we can do everything we need without it. For a development incorporating the notion of proper position, see [**Wag11**].

5.1. Equivalences. Recall from Proposition 1.3 the relation between stability and hyperbolicity: a real homogeneous polynomial is stable if and only if it has a cone of hyperbolicity containing the (strictly) positive orthant. A number of equivalent formulations of stability will prove useful. If $f \in \mathbb{C}[z_1, \dots, z_d]$ is any polynomial, let m denote the maximum total degree of any monomial in f and define the homogenization f_H of f to be the unique degree-m homogeneous polynomial in the variables $z_1, \dots, z_{d+1}$ such that $f_H(z_1, \dots, z_d, 1) = f(z_1, \dots, z_d)$. We may write explicitly $f_H(z_1, \dots, z_{d+1}) = z_{d+1}^m f(z_1/z_{d+1}, \dots, z_d/z_{d+1})$. The following proposition is proved in [**BBL09**, Section 4].

PROPOSITION 5.3 (homogenization). *The real polynomial $f \in \mathbb{R}[z_1, \dots, z_d]$ is stable if and only if the homogeneous polynomial f_H is hyperbolic in all directions $\mathbf{v}$ such that $v_j > 0$ for $1 \le j \le d$ and $v_{d+1} = 0$.*

If f also has nonnegative coefficients then the four properties are all equivalent:

(i) f is stable;
(ii) f_H is stable;

(iii) f_H *is hyperbolic with respect to some vector in the nonnegative orthant;*
(iv) f_H *is hyperbolic with respect to every vector in the positive orthant.*

PROOF: For one direction, assume f_H to be hyperbolic in every direction $\mathbf{y}$ with y_j positive when $j \leq d$ and $y_{d+1} = 0$. For any $\mathbf{x} \in \mathbb{R}^d$ and $\mathbf{y} \in \mathbb{R}^d_+$, the polynomial

$$t \mapsto f(\mathbf{x} + t\mathbf{y}) = f_H((\mathbf{x}, 1) + t(\mathbf{y}, 0))$$

is not identically zero because $\lim_{t\to\infty} t^{-m} f(\mathbf{x}+t\mathbf{y})$ is equal to $f_H(\mathbf{y}, 0) \neq 0$. It has only real zeros, by hyperbolicity of f_H in direction $(\mathbf{y}, 0)$, hence i is not a root, hence $f(\mathbf{x}+i\mathbf{y}) \neq 0$; but $\mathbf{x} \in \mathbb{R}^d$ and $\mathbf{y} \in \mathbb{R}^d_+$ are arbitrary, so this is stability of f.

For the other direction, suppose f_H fails to be hyperbolic in a direction $\mathbf{v}$ for which $v_j > 0, 1 \leq j \leq d$, and $v_{d+1} = 0$. Then $f_H(\mathbf{x} + t\mathbf{v})$ has a non-real zero $t = u$ for some $\mathbf{x} \in \mathbb{R}^{d+1}$. The complex conjugate of u is also a zero so we may assume without loss of generality that the imaginary part of u is positive. The vector $\mathbf{y} := \mathbf{x} + u\mathbf{v}$ lies in $\mathcal{H}^d \times \mathbb{R}$. There are three cases. First, if $y_{d+1} > 0$ then dividing though by this gives a zero of f in $\mathcal{H}^d$, implying f is not stable. Second, if $y_{d+1} < 0$, dividing yields a zero in $-\mathcal{H}^d$ and conjugating, we see again that f is not stable. Finally, if $y_{d+1} = 0$, we may use perturbation to reduce to the case where $y_{d+1} \neq 0$.

For the equivalence of the four properties in the case of nonnegative coefficients, recall that the cone of hyperbolicity of f_H containing some vector $\mathbf{v}$ is the connected component in $\mathbb{R}^{d+1}$ that contains $\mathbf{v}$ of the set where f_H is nonzero. By nonnegativity of the coefficients, if this cone contains any vector in the closed nonnegative orthant then it contains all vectors in the positive orthant. This establishes the equivalence of (iii) and (iv). Equivalence between (iii) and (ii) is Proposition 1.3. Finally, we note that (ii) implies (i) by setting $z_{d+1} = 1$ (as is allowed by the last consequence of Lemma 5.2), while (i) implies (iii) by the first part of this proposition. □

REMARK. The perturbation argument also proves that stability of f implies stability of $\mathsf{hom}(f)$, where $\mathsf{hom}(f)(z_1, \ldots, z_d) := f_H(z_1, \ldots, z_d, 0)$ is the leading homogeneous part of f; see, e.g., [**COSW04**, Proposition 2.2].

The special case of multi-affine polynomials. Suppose f is multi-affine, that is, no variable appears in any monomial with power two or higher. The following equivalence is taken by some to be the definition of stability in the multi-affine case. Its significance will become evident in the next section.

THEOREM 5.4 ([**Brä07**, Theorem 5.6]). *Let f be real and multi-affine. Then f is stable if and only if the inequality*

$$\frac{\partial f}{\partial x_i}(\mathbf{x}) \frac{\partial f}{\partial x_j}(\mathbf{x}) \geq f(\mathbf{x}) \frac{\partial^2 f}{\partial x_i \partial x_j}(\mathbf{x}) \tag{5.1}$$

holds for all real $\mathbf{x}$ and distinct $i, j \leq n$.

The proof given here is considerably simpler than the published proof, which is a "proper position" argument. It comes from the same source (P. Brändén, personal communication) and begins with the following lemma.

LEMMA 5.5. *Let* $Q, R \in \mathbb{C}[z_1, \ldots, z_n]$ *and define* $P = Q + z_{n+1} R \in \mathbb{C}[z_1, \ldots, z_{n+1}]$. *Let* $\Omega \subseteq \mathbb{C}^n$ *be any connected set and let* D_1, D_2 *be closed subsets of* $\mathbb{C}$ *with disjoint interiors, such that* $D_1 \cup D_2 = \mathbb{C}$ *and* $J := D_1 \cap D_2$ *a simple curve separating* D_1 *and* D_2.

If P *has no roots in* $\Omega \times J$ *and* R *has no roots in* Ω *then* P *either has no roots in* $\Omega \times D_1$ *or has no roots in* $\Omega \times D_2$.

PROOF: P having no roots in $\Omega \times J$ is equivalent to $-Q(\mathbf{z})/R(\mathbf{z}) \notin J$ for all $\mathbf{z} \in \Omega$. Hence $-Q/R$ maps Ω either to the interior of D_1 or the interior of D_2, hence to D_2^c or D_1^c, and the result follows. □

PROOF OF THEOREM: For the forward implication, suppose f is stable. Let e_i denote the ith standard basis vector and define $G(s,t) := f(\mathbf{x} + te_i + se_j)$. For any $\mathbf{x} \in \mathbb{R}^d$, the function G is stable in the variables (s,t). To see this, note that this is equivalent to f having no zeros in $\overline{\mathcal{H}}^d$ with coordinates i and j in $\mathcal{H}$; this is ruled out by Lemma 5.2. By multi-affinity, we may express G as $a + bs + ct + dst$ where a, b, c, d are given by partial derivatives:

$$G(s,t) = f(\mathbf{x}) + \frac{\partial f}{\partial z_i}(\mathbf{x})s + \frac{\partial f}{\partial z_j}(\mathbf{x})t + \frac{\partial^2 f}{\partial z_i \partial z_j}(\mathbf{x})st\,.$$

For bivariate multi-affine polynomials $a + bs + ct + dst$, stability is equivalent to $ad \le bc$, and the forward direction follows.

For the reverse direction, use induction on n. When $n = 1$ the result is true because all nonzero real polynomials $a + bx$ are stable. Assume the reverse result for n and suppose $P = Q + z_{n+1}R$ satisfies (5.1). If Q or R is identically zero then we are done by induction, so assume not. For any real x, it is clear that $P(z_1, \ldots, z_n, x)$ satisfies (5.1), hence, by induction, the polynomials Q, R and $Q + xR$ are all stable or identically zero. If $Q + \alpha R \equiv 0$ for some real α then $P = (z_{n+1} - \alpha)R$ which is stable. We may assume, therefore, that there is no α for which $Q + \alpha R$ is identically zero. It follows that P has no zeros in $H^n \times \mathbb{R}$ where H is the open upper half-plane. By Lemma 5.5 this implies that either P or $P(z_1, \ldots, z_n, -z_{n+1})$ is stable. In the latter case, by the forward direction, equation (5.1) holds with the signs reversed for $j = n+1$ and $i \le n$, and hence with equality for such i and j. This works out to $(\partial R/\partial x_i)Q = (\partial Q/\partial x_i)R$ for all $i \le n$, that is, Q and R are multiples of each other, which again implies that P is stable. □

5.2. Operations preserving stability. In a brilliant series of papers [**BB09a, BB09b, BB10**], Borcea and Brändén finished off the problem of extending the Pólya-Schur Theorem to multivariate stability preserving maps. This work not only generalizes from one variable to several, but also from multiplier sequences to arbitrary $\mathbb{C}$-linear maps. I will present here two of their results, one for multiplier sequences and one giving universal test functions for general linear maps, proving only those parts that will be of use later.

As before, if $\boldsymbol{\lambda} = \{\lambda_{\mathbf{r}} : \mathbf{r} \in \mathbb{Z}_+^d\}$ is an array of real numbers, we denote by $T_{\boldsymbol{\lambda}}$ the operator for which

$$T_{\boldsymbol{\lambda}} \left(\sum_{\mathbf{r}} a_{\mathbf{r}} \mathbf{z}^{\mathbf{r}} \right) = \sum_{\mathbf{r}} \lambda_{\mathbf{r}} a_{\mathbf{r}} \mathbf{z}^{\mathbf{r}} .$$

Also generalizing from before, we say that $\boldsymbol{\lambda}$ is a **multivariate multiplier sequence** if $T_{\boldsymbol{\lambda}}$ preserves the class of real stable polynomials. The following theorem is proved in [**BB10**, Theorem 1.8].

THEOREM 5.6 (multivariate multiplier sequences). *The array $\boldsymbol{\lambda}$ is a d-variate multiplier sequence if and only if there are d univariate multiplier sequences $\boldsymbol{\lambda}^{(1)}, \ldots, \boldsymbol{\lambda}^{(d)}$ such that*

$$\lambda_{\mathbf{r}} = \lambda^{(1)}_{r_1} \cdots \lambda^{(d)}_{r_d}$$

and satisfying a further sign condition: either every $\lambda_{\mathbf{r}}$ is nonnegative, or every $\lambda_{\mathbf{r}}$ is nonpositive, or the same holds for $(-1)^{|\mathbf{r}|}\boldsymbol{\lambda}$.

PROOF IN ONE DIRECTION: The easy direction, and all we will need below, is to see that the product of univariate nonnegative multiplier sequences is a multiplier sequence (also that nonnegative multiplier sequences preserve complex stability, not just real stability). Let $\lambda_{\mathbf{r}} = \prod_{j=1}^d \lambda^{(j)}_{r_j}$ where each sequence $\boldsymbol{\lambda}^{(j)}$ is a univariate multiplier sequence. First assume that $\lambda^{(j)}$ is identically 1 for $j \geq 2$. Let f be a real stable polynomial of d variables. Fixing $z_2, \ldots, z_d$ in the upper half-plane, the polynomial $f_1 := f(\cdot, z_2, \ldots, z_d)$ is univariate stable, and by the stability preserving assumption on $\boldsymbol{\lambda}^{(i)}$, we see that $T_{\boldsymbol{\lambda}_1}(f_1)$ is univariate stable. This polynomial having no zeros in the upper half-plane is equivalent to $T_{\boldsymbol{\lambda}} f$ having no zeros with z_1 in the upper half-plane and the specified values of $z_2, \ldots, z_d$. Because these values were arbitrary values in the upper half-plane, this finishes the proof in this special case. For the general case, write $T_{\boldsymbol{\lambda}}$ as the composition of operators of this form in each coordinate. □

One consequence of this is that the multi-affine part of a stable polynomial is stable.

COROLLARY 5.7. *Let f be a stable polynomial in d variables and let $f_{\rm ma}$ denote the multi-affine part of f, that is, the sum of the square-free monomials in f. Then $f_{\rm ma}$ is stable.*

PROOF: The sequence $a_0 = 1, a_1 = 1, a_n = 0$ for $n \geq 2$ is a multiplier sequence. By Theorem 5.6, the multivariate sequence defined by $\boldsymbol{\lambda}_{\mathbf{r}} = 1$ if $\mathbf{r}_j = 0$ or 1 for all $j \leq d$ and $\boldsymbol{\lambda}_{\mathbf{r}} = 0$ otherwise is a multiplier sequence. This multiplier sequence extracts the multi-affine part. □

A complete characterization of linear operators preserving stability of complex polynomials was proved in [**BB09a**, Theorem 2.3]. We will discuss here only the version of this result concerning operators preserving stability of multi-affine functions. We need to define what Borcea and Brändén call the **algebraic symbol** G_T of a linear operator T. Their definition requires a degree bound on each variable. We consider here only the multi-affine case; if you compare to the original text, you should set the degree bound vector κ to $(1, \ldots, 1)$. Let $\mathbb{C}_{\mathrm{ma}}[z_1, \ldots, z_d]$ denote the space of complex multiaffine polynomials in d variables. For a linear operator $T :$ $\mathbb{C}_{\mathrm{ma}}[z_1, \ldots, z_d] \to \mathbb{C}[z_1, \ldots, z_d]$ define its algebraic symbol to be the element $G_T \in \mathbb{C}[z_1, \ldots, z_d, w_1, \ldots, w_d]$ satisfying

$$G_T(z_1, \ldots, z_d, w_1, \ldots, w_d) = \sum_{S \subseteq \{1,\ldots,d\}} \prod_{j \notin S} w_j T\left(\prod_{j \in S} z_j\right) . \tag{5.2}$$

In other words, G applies T to $\prod_{j=1}^d (z_j + w_j)$ by treating $\{w_j\}$ as constants and applying T to the resulting monomials in $\{z_j\}$.

THEOREM 5.8 ([**BB09a**, Lemma 3.2]). *Let* $T : \mathbb{C}_{\mathrm{ma}}[z_1, \ldots, z_d] \to$ $C[z_1, \ldots, z_d]$ *be any* $\mathbb{C}$*-linear operator. Then* T *preserves stability if and only one of the two following conditions hold:*

(a) There is a stable polynomial P *and a linear functional* $\alpha : \mathbb{C}_{\mathrm{ma}}[z_1, \ldots, z_d]$ $\to \mathbb{C}$ *such that* $T(f) = \alpha(f)P$*;*

(b) The polynomial $G_T(z_1, \ldots, z_d, w_1, \ldots, w_d)$ *is stable as a complex polynomial in* $2d$ *variables.*

REMARK. This is a very powerful theorem and it is worth asking what part of it one needs to understand in order to understand the Borcea–Brändén–Liggett theory of negatively dependent random variables surveyed in Section 6. The answer is that we only use one direction (sufficiency) and this is only to prove Proposition 6.17, which could be proved elementarily as it concerns only bivariate functions. Nevertheless, because this would be messy and unenlightening, we will prove the sufficiency direction of Theorem 5.8 here, which we will then use to prove Proposition 6.17. The proof relies on a lemma proved thirty years ago by Lieb and Sokal [**LS81**], which we will not prove here (the proof is half a page in [**Wag11**]).

LEMMA 5.9 ([**LS81**, Lemma 2.3]). *Let* $P(\mathbf{z}) + wQ(\mathbf{z})$ *be stable in* $\mathbb{C}[z_1, \ldots, z_d, w]$*. If the degree of the variable* z_j *is at most 1 then the polynomial*

$$P(\mathbf{z}) - \frac{\partial Q(\mathbf{z})}{\partial z_j}$$

is either identically zero or is stable. □

PROOF OF SUFFICIENCY: If T satisfies (a) the result is immediate, so assume T satisfies (b). Because $w \mapsto -w^{-1}$ preserves $\mathcal{H}$, we see that $G_T(\mathbf{z}, \mathbf{w})$ is stable if and only if $w_1 \cdots w_d G_T(\mathbf{z}, -\mathbf{w}^{-1})$ is stable; here $-\mathbf{w}^{-1}$ denotes

$(-w_1^{-1}, \ldots, -w_d^{-1})$. From this, we see that if $f \in \mathbb{C}[v_1, \ldots, v_d]$ is stable and multi-affine, then

$$(5.3) \qquad (w_1 \cdots w_d) \cdot G_T(\mathbf{z}, -\mathbf{w}^{-1}) \cdot f(\mathbf{v}) = \sum_{S \subseteq \{1,\ldots,n\}} T[\mathbf{z}^S](-\mathbf{w})^S f(\mathbf{v})$$

is stable in $\mathbb{C}[\mathbf{z}, \mathbf{w}, \mathbf{v}]$. We now apply the Lieb-Sokal lemma d times: the jth time we take $w = w_j$ and $z_j = v_j$. Each time, the Lieb Sokal lemma replaces w_j in (5.3) by $-(\partial/\partial v_j)$. The final result is that

$$(5.4) \qquad \sum_{S \subseteq \{1,\ldots,n\}} T[\mathbf{z}^S] f^{(S)}(\mathbf{v})$$

is stable in $\mathbb{C}[\mathbf{z}, \mathbf{v}]$. Specializing each v_j to zero preserves stability. But this specialization yields $\sum_S f^{(S)}(0) T[z^S]$, which is another name for $Tf(\mathbf{z})$. Thus stability of f implies stability of Tf, proving the theorem. □

5.3. More closure properties. The multivariate Pólya-Schur theorems of Borcea and Brändén are powerful but there are yet more closure properties that are useful in probabilistic settings because they have specific probabilistic meanings. We state these now but defer the proofs to the next section.

Let $n_j, 1 \leq j \leq d$ be the maximum degree of the variable z_j in a given polynomial $f \in \mathbb{C}[z_1, \ldots, z_d]$. One natural way to create a multi-affine polynomials out of f is to replace each occurrence of each power z_j^r by $\binom{n_j}{r}^{-1}$ $e_r(z_{j,1}, \ldots, z_{j,n_j})$, where e_r is the rth elementary symmetric function. Here $\{z_{j,s} : 1 \leq j \leq d, 1 \leq s \leq n_j\}$ are a collection of **clone** variables. Formally, define the **polarization** f_p of f to be the unique polynomial symmetric separately in each set of clone variables $z_{j,1}, \ldots, z_{j,n_j}$ such that substituting the value z_j for all the clones $z_{j,s}$ yields the original function f. In Section 6.4 we will show that stability is closed under polarization.

One final useful closure result is the following. The proof will be given after a probabilistic interpretation is given in Section 6.4.3.

THEOREM 5.10 (partial symmetrization). *Let f be a d-variate complex poynomial. Fix $1 \leq i < j \leq d$ and define τf to be f with the roles of z_i and z_j swapped:*

$$\tau f(z_1, \ldots, z_d) = f(z_1, \ldots z_{i-1}, z_j, z_{i+1}, \ldots, z_{j-1}, z_i, z_{j+1}, \ldots, z_d).$$

Fix $\theta \in [0,1]$ and define $f_{\theta;i,j} := (1-\theta)f + \theta\tau(f)$. If f is stable, so is $f_{\theta;i,j}$.

6. Negative dependence

6.1. A brief history of negative dependence. When random variables in a collection influence each other in a consistent direction, this can be very useful for obtaining one-sided bounds on moments and probabilities of natural events. For example, when all pairwise correlations are negative, the variance of the sum is bounded above by the sum of the variances. For this reason, the literature is sprinkled with definitions of various properties of positive and negative dependence, that hold in examples of interest and which imply useful consequences. Some of these apply to random variables taking values in $\mathbb{R}, \mathbb{R}^+, \mathbb{Z}$ or $\mathbb{Z}^+$. To keep matters simple and coherent, we will discuss only the case where the variables are **binary valued**, that is, take values in the set $\{0,1\}$. The joint law of d binary variables is a probability measure on $\mathcal{B}_d := \{0,1\}^d$.

The simplest and one of the weakest conditions is pairwise positive (respectively negative) correlation: $\mathbb{E}X_iX_j \geq (\mathbb{E}X_i)(\mathbb{E}X_j)$ (respectively $\leq$), which may also be written as $\mathbb{P}(X_i = X_j = 1) \geq \mathbb{P}(X_i = 1)\mathbb{P}(X_j = 1)$ (respectively $\leq$). As we have just seen, this condition is strong enough to imply one-sided bounds on the variance of $\sum_j X_j$. It is not strong enough to give bounds on the moments. For this, one requires at least positive or negative **cylinder dependence**. Positive cylinder dependence is said to hold if

$$\mathbb{E}\prod_{j\in S} X_j \geq \prod_{j\in S} \mathbb{E}X_j \tag{6.1}$$

for all subsets $S \subseteq \{1,\ldots,n\}$. Negative cylinder dependence is the reverse inequality. Sometimes this is strengthened to require as well that

$$\mathbb{E}\prod_{j\in S}(1-X_j) \geq \prod_{j\in S}(1-\mathbb{E}X_j)\,. \tag{6.2}$$

The equation (6.1) immediately implies that the moments of $Z := \sum_j X_j$ are at least what they would be for independent Bernoullis with the same marginals, while negative cylinder dependence implies the moments are at most what they would be in the independent case. Expanding the moment generating function $\phi(\lambda) := \mathbb{E}e^{\lambda Z}$ in powers of Z, and then into monomials in the variables X_j shows that positive (respectively negative) cylinder dependence implies that for $\lambda \geq 0$, the moment generating function $\phi(\lambda)$ is at least (respectively at most) what it would be for independent Bernoullis of the same marginal; including the inequality for complementary cylinders (6.2) extends the inequality for moment generating functions to $\lambda < 0$. The bounds on moment generating functions transfer to concentration inequalities (see, e.g., [**PS97**, Theorem3.4]).

A yet stronger property is association. Say that the collection of random variables $\{X_1,\ldots,X_d\}$ is **positively associated** if

$$\mathbb{E}f(X_1,\ldots,X_d)g(X_1,\ldots,X_d) \geq \mathbb{E}f(X_1,\ldots,X_d)(\mathbb{E}g(X_1,\ldots,X_d)$$

whenever $f,g : \mathcal{B}_d \to \mathbb{R}$ are nondecreasing functions (with respect to the coordinatewise partial order on $\mathcal{B}_d$). Negative association cannot be defined by simply reversing the inequality because any function f is always nonnegatively correlated with itself. Therefore, we amend the definition to say that $\{X_1,\ldots,X_d\}$ are **negatively associated** if $\mathbb{E}fg \leq \mathbb{E}f\,\mathbb{E}g$ whenever $f,g : \mathcal{B}_d \to \mathbb{R}$ are nondecreasing and there is a set $S \subseteq \{1,\ldots,n\}$ such that f depends only on $\{X_j : j \in S\}$ while g depends only on $\{X_j : j \notin S\}$.

One of the most useful results on positive association is a result of Fortuin, Kasteleyn and Ginibre [**FKG71**]. Say that y covers z in the Boolean lattice $\mathcal{B}_d$ if $y > z$ in the coordinatewise partial order but there is no element strictly between y and z; we denote this relation by

$$y \gtrdot z \, ,$$

If x and y both cover z and w covers x and y, we call this configuration a face of the lattice $\mathcal{B}_d$. Say that the probability measure μ satisfies the **positive lattice condition** if whenever x, y, z, w form a face of $\mathcal{B}_d$ (with z and w at the bottom and top), $\mu(x)\mu(y) \leq \mu(z)\mu(w)$. It is immediate to see that this property for faces implies $\mu(x \wedge y)\mu(x \vee y) \geq \mu(x)\mu(y)$ for every pair x, y, hence we also call in (multiplicative) submodularity. What is not so obvious is the theorem now known as the FKG Theorem.

THEOREM 6.1 (FKG). *If μ satisfies the positive lattice condition then μ is positively associated.* □

The power of this theorem is that its hypotheses are easy to check while its conclusions are quite strong. In many statistical mechanical systems, ratios of probabilities over a face of $\mathcal{B}_d$ are easy to compute even if the absolute probabilities are not. The FKG theorem immediately implies positive association for ferro-magnetic Ising models, the random cluster model with $q \geq 1$, and a number of other models from physics, many of which are surveyed in [**Lig85**].

For negative dependence the situation is not as nice. The negative lattice condition (reverse the inequality on each face) does not imply negative association, or seemingly anything of value. Unlike the positive lattice condition it is not closed under the most natural of operations, namely integrating out one variable (probabilistically this means ignoring this variable and viewing μ as a measure on $\mathcal{B}_{n-1}$). As a result, it has been difficult to identify negatively dependent laws and to establish properties such as negative association.

6.2. Search for a theory. For some time now, a more satisfying theory of negative dependence has been sought. The goal was to find a property of laws μ on $\mathcal{B}_d$ which would be:

- checkable for a handful of known or conjectured **examples**;
- shown to imply **consequences** such as negative association;
- closed under natural **operations**.

This goal was popularized in the article [**Pem00**] which posed several challenges and stated many conjectures but contained few concrete results. **Spoiler alert:** for binary valued random variables, the property is that the probability generating function is stable; such a probability distribution is called *strong Rayleigh*; see Theorem 6.9 below. In the forthcoming lists of examples, consequences and closure properties, only in two cases are they not known to hold for the class of strong Rayleigh measures: Example 6.4

(the random cluster measure) is conjectured to be Rayleigh but known not to be strong Rayleigh, and the proposed closure property (d) from Section 6.2.3 (restriction to an interval) is known to fail.

One open question remaining from [**Pem00**] is whether the property known there as h-NLC$^+$ implies negative association. To define h-NLC$^+$, let us first define an **external field**. Let μ be a probability measure on $\mathcal{B}_n$ and let $\boldsymbol{\lambda}_j > 0$ be positive real numbers, for $1 \leq j \leq d$. Define a measure μ' by

$$\mu'(x_1, \ldots, x_d) = \frac{\prod_{j=1}^d \lambda_j^{x_j}}{Z} \mu(x_1, \ldots, x_d)$$

where $Z := \sum_{\mathbf{x}} \boldsymbol{\lambda}^{\mathbf{x}} \mu(\mathbf{x})$ is the normalizing constant. We call μ' the perturbation of μ by the external field $\boldsymbol{\lambda}$. The negative lattice condition is not closed under integrating out a variable; we say that the hereditary negative lattice condition plus external fields holds (h-NLC$^+$) if the negative lattice condition holds for all measures obtained from μ by imposing an external field and integrating out some of the variables. This was later shown to be equivalent to Wagner's **Rayleigh condition** [**Wag08**], which will discuss further in Section 6.3. We now survey the proposed examples, consequences and closure properties for the desired class of negatively dependent laws.

6.2.1. *Proposed examples.*

EXAMPLE 6.2 (conditioned Bernoullis). *Let μ be the law of independent Bernoulli variables $\{X_1, \ldots, X_d\}$ with means $p_1, \ldots, p_d$, let k be an integer between 1 and d and let $\mu' := (\mu | \sum_{j=1}^d X_j = k)$ be the conditional law of $\{X_j\}$ given that they sum to k. This should be negatively dependent. Note that this is equal to the uniform measure on k-subsets of $\{1, \ldots, n\}$ under the external field $\lambda_j = p_j/(1-p_j)$.*

EXAMPLE 6.3 (spanning trees). *Let $G = (V, E)$ be a finite connected simple graph and let $\{\lambda_e : e \in E\}$ be nonnegative weights. Let W be the set of subsets $T \subseteq E$ such that the resulting (V, T) is connected and has no cycles. Each $T \in W$ will have cardinality exactly $|V| - 1$. Such subgraphs are called **spanning trees** of G. Let $w(T) := \prod_{e \in T} \lambda_e$ denote the weight of T under the multiplicative weighting scheme determined by $\boldsymbol{\lambda}$. We define the weighted spanning measure μ (depending on G and $\boldsymbol{\lambda}$) by*

$$\mu(T) = \frac{w(T)}{\sum_{T \in W} w(T)} .$$

This was proven in [**Pem91**] *to have negative correlations and in* [**FM92**] *to have negative association, and we would hope it to included under any reasonable definition of "negatively dependent".*

EXAMPLE 6.4 (random cluster measure). *The random cluster model on a graph $G = (V, E)$ with parameters $q > 0$ and $\{\lambda_e : e \in E\}$ is the probability* *measure* μ *on* $\{0,1\}^E$ *obtained by normalizing the weights*

$$w(T) = \prod_e \lambda_e^{T(e)} q^{N(T)}$$

where $T \subseteq E$ *and* $N(T)$ *is the number of connected components of the subgraph* (V,T) *of* G. *When* $q=2$ *this is equivalent to the Ising model for ferromagnetism and for integers* $q \geq 2$ *it is equivalent to the Potts model, also from statistical physics (see, e.g.* [**Wu88**]*). When* $q \geq 1$, *the FKG Theorem immediately shows* μ *to be positively associated. When* $0 < q < 1$, *the measure is conjectured but not known to be negatively associated. We would hope that the right definition of negative dependence would settle this conjecture.*

EXAMPLE 6.5 (exclusion measures). *The* ***exclusion process*** *is a continuous time Markov chain on* $\mathcal{B}_d$ *as follows. The state* $\mathbf{x} \in B_d$ *is interpreted as a collection of particles at the sites* $j \in \{1,\ldots,d\}$ *for which* $x_j = 1$; *the sites* j *for which* $x_j = 0$ *are considered vacant. Let* $\lambda_{\{i,j\}}$ *be nonnegative real numbers, and independently at rate* $\lambda_{\{i,j\}}$, *let the values* x_i *and* x_j *swap. This is interpreted as a particle at one site* i *or* j *jumping to the other site. If both sites or neither site is occupied, then nothing happens. The exclusion process is a special case of the more general* ***exchange process***, *where the labels of the sites are arbitrary, and in particular might be distinct rather than being drawn from the two element set* $\{0,1\}$. *Let* μ_t *be the law of the exclusion process at time* t, *starting from some deterministic state* η *at time 0. It was conjectured that* μ_t *is always negatively associated. We would like the definition of negatively dependent measures to include* μ.

EXAMPLE 6.6 (determinantal measures with Hermitian kernels). *Let* K *be a* $d \times d$ *Hermitian matrix with spectrum in* $[0,1]$. *For* $S \subseteq \{1,\ldots,d\}$, *let* $w(S)$ *denote the determinant of the submatrix* K_S *of* K *when restricted to rows and columns in the set* S. *There is a unique probability measure* μ *on subsets of* $\{1,\ldots,d\}$ *such that*

$$\mu\{T : T \supseteq S\} = |K_S|$$

for each S. *This is called the* ***determinantal*** *measure with kernel* K. *Such measures were proved in* [**Lyo03**] *to be negatively associated. We would hope these measures satifsy our definition of negative dependence.*

6.2.2. *Proposed consequences.* It was proposed in [**Pem00**] that the right definition of negative dependence for binary variables would imply the following properties, with definitions immediately to follow.

(a) negative association
(b) stochastically increasing levels
(c) stochastic covering property
(d) log-concave rank sequence

Recall that a probability measure μ on a lattice W is said to **stochastically dominate** a law ν, written $\mu \succeq \nu$, if $\mu(A) \geq \nu(A)$ for every upwardly closed set A (the set A is upwardly closed if $x \in A$ and $y > x$ implies $y \in A$). An equivalent condition is stated in terms of coupling: $\mu \succeq \nu$ if and only if there is a measure Q on $W \times W$ whose first marginal is μ, whose second marginal is ν, and which is supported on $\{(x, y) \in W^2 : x \geq y\}$. In other words, we may simultaneously sample from μ and ν in such a way that the sample from μ is always greater than or equal to the sample from ν.

Given a measure μ on $\mathcal{B}_d$ and an event $G \subseteq \mathcal{B}_d$ with nonzero measure, denote by $(\mu|G)$ the measure $\mu(\cdot)/\mu(\cdot \cap G)$ obtained by conditioning on G. The conditional measure $(\mu|X_j = 1)$ may be identified with a measure on $\mathcal{B}_{d-1}$ by ignoring the jth coordinate.

PROPERTY 6.1 (stochastically increasing levels). *Say that a measure μ on $\mathcal{B}_d$ has stochastically increasing levels if $(\mu|\sum_j x_j = k) \preceq (\mu|\sum_j x_j = k+1)$ for each k such that $\mu(\sum_j x_j = k)$ and $\mu(\sum_j x_j = k+1)$ are both nonzero.*

The property of having stochastically increasing levels is not implied by negative association, as was hoped in [**Pem00**], but is a desired consequence of the right definition of negative dependence.

PROPERTY 6.2 (stochastic covering). *Say that a measure ν on a lattice W stochastically covers the measure μ, and denote this by $\nu \rhd \mu$, if there is a coupling measure Q on W^2 with marginals ν and μ, such that Q is supported on the set $\{(x, y) \in W^2 : x = y \text{ or } x \gtrdot y\}$. In other words, ν stochastically dominates μ but "only by one", in the sense that you can sample from μ by sampling from ν and then flipping at most one one to a zero.*

Negative association of a measure μ implies that $(\mu|X_j = 0) \succeq (\mu|X_j = 1)$ when viewed as a measure on $\mathcal{B}_{d-1}$. The **stochastic covering property** is defined in [**PP11**] to hold when $(\mu|X_j = 0) \rhd (\mu|X_j = 1)$, viewed as measures on $\mathcal{B}_{d-1}$. This property was wrongly conjectured in [**Pem00**] to follow from negative association. We would hope for it to follow from the right definition of negative dependence.

PROPERTY 6.3 ((ultra) log-concave rank sequence). *The **rank sequence** for the measure μ on $\mathcal{B}_d$ is the sequence $(a_k := \mu\{\sum_j X_j = k\})_{0 \leq k \leq d}$. It was wrongly conjectured to follow from negative association that the rank sequence is log-concave. The right definition of negative dependence turns out to imply not only log-concavity but **ultra log-concavity**, namely that for $0 \leq k \leq d$, equation (4.4) holds, which I repeat here for convenience.*

$$\frac{a_k^2}{\binom{d}{k}^2} \geq \frac{a_{k-1}}{\binom{d}{k-1}} \frac{a_{k+1}}{\binom{d}{k+1}} . \tag{6.3}$$

6.2.3. *Proposed closure properties.* In addition to negative association, it was thought that the right negative dependence property would imply the following closure properties.

(a) closure under products, projections (integrating out a variable), and limits;
(b) closure under external fields;
(c) conditioning on the event $\{\sum_j X_j = k\}$;
(d) more generally, conditioning on the event $k_1 \le \sum_j X_j \le k_2$;
(e) total symmetrization: replacing $\mu(\mathbf{x})$ by $(1/n!)\sum_{\pi\in S_d}\mu(\pi\mathbf{x})$ where π acts by permuting the coordinates;
(f) more generally, evolution via the symmetric exclusion process for any finite time.

It turns out that (c) is too strong, but that the class of strong Rayleigh measures is indeed closed under **rank rescaling**, (see the upcoming definition) when the rank sequence $\mathbf{b}$ is ultra log-concave.

DEFINITION 6.7 (rank rescaling). *Let μ be a measure on $\mathcal{B}_d$ and let $\mathbf{b} := (b_0,\ldots,b_d)$ be a vector of nonnegative real numbers. For $\mathbf{x} = (x_1,\ldots,x_d) \in \mathcal{B}_d$ let $N(\mathbf{x}) := \sum_{j=1}^d x_j$ denote the sum of the coordinates. For nontriviality assume that $Z := \sum_{i=0}^d b_i\mu\{N = i\} > 0$. The rank rescaling of μ by $\mathbf{b}$ is the measure $\mu_{\mathbf{b}}$ defined by*

$$\mu_{\mathbf{b}}(\mathbf{x}) = \frac{N(\mathbf{x})}{Z}\,\mu(\mathbf{x})\,.$$

6.3. The grail is found: application of stability theory to joint laws of binary random variables. It is now acknowledged that the "correct" definition of negative dependence for binary random variables is the following definition due to [**BBL09**].

DEFINITION 6.8 (strong Rayleigh). *Say that the probability measure μ on $\mathcal{B}_d$ is strongly Rayleigh (also "strong Rayleigh") if its generating polynomial $f \in \mathbb{C}[z_1,\ldots,z_d]$ defined by*

$$f(\mathbf{z}) := \sum_{\omega\in\mathcal{B}_d} \mu(\omega)\mathbf{z}^\omega$$

is a real stable polynomial.

As a generating polynomial, f is automatically multi-affine, with real nonnegative coefficients. Recalling Theorem 5.4, we see that strong Rayleigh is equivalent to the inequality (5.1) on the mixed partial derivatives of f holding for all real $\mathbf{x}$. This is sometimes taken as the definition of strong Rayleigh.

The origin of the name comes from Wagner's Rayleigh property. This is the mixed partial inequality (5.1) but only required to hold for nonnegative real arguments. As was mentioned before, this is equivalent to h-NLC$^+$, which is not hard to believe because varying $\mathbf{x}$ over $\mathbb{R}_+^d$ corresponds to varying the external field arbitrarily. Strangely, when this is required to hold for negative values of the external field as well, the property becomes more natural and robust. Because the Rayleigh property is equivalent to h-NLC$^+$ and is obviously a consequence of the strong Rayleigh property, we see immediately that strong Rayleigh implies pairwise negative correlations. This is, of course, just the tip of the iceberg.

THEOREM 6.9 (properties of strong Rayleigh measures).
(i) Strong Rayleigh measures are negatively associated.

(ii) Strong Rayleigh measures have stochastically increasing levels, satisfy the stochastic covering property 6.2, and have ultra log-concave rank sequences as in (6.3). The class of strong Rayleigh measures is also closed under rank rescaling by sequences $(b_0, \ldots, b_d)$ *when the nonzero values* b_i *are the coefficient sequence of a stable univariate polynomial* $\sum_{i=0}^d b_i z^i$*. In particular, this implies closure of the class of strong Rayleigh measures under conditioning on* $\{\sum_j X_j = k\}$ *or* $\{k \le \sum_j X_j \le k+1\}$*.*

(iii) The class of strong Rayleigh measures is closed under products, projections, external fields, total symmetrization and the symmetric exclusion dynamics.

As an example of the utility of this theorem, here is one of the two applications that brought the theory to my attention.

EXAMPLE 6.10 (concentration inequalities for Lipschitz functions). *Let* μ *be a measure on* $\mathcal{B}_d$ *and let* $f : \mathcal{B}_d \to \mathbb{R}$ *be Lipschitz with constant 1. When* μ *is a product measure, the martingale* $W_k := \mathbb{E}(f \mid \omega_1, \ldots, \omega_k)$ *has bounded increments* $|W_k - W_{k-1}| \le 1$*. It follows from Azuma's inequality (see, e.g,* [**AS08**]*) that* f *satisfies the Gaussian concentration inequality*

$$\mathbb{P}(f - \mathbb{E}f \ge a) \le e^{-a^2/(2n)} . \tag{6.4}$$

In fact independence is not needed. The stochastic covering property was defined in [**PP11**] *precisely to imply* $|W_k - W_{k-1}| \le 1$*. It follows that* (6.4) *holds for any strong Rayleigh measure.*

6.4. Proof of Theorem 6.9.

6.4.1. *Proof of part (i) : negative association.* "Strong Rayleigh implies negative association" was the result of Borcea–Brändén–Liggett that pointed the way to application of stable polynomial theory to the probabilistic setting. We follow their proof, which begins by paving the way for a reduction to the case of homogeneous measures.

LEMMA 6.11 (homogenization). *Let* $\{X_1, \ldots, X_d\}$ *be a collection of random variables whose law is strong Rayleigh and define* $X_{d+1} := d - \sum_{j=1}^d X_j$*. Then the collection* $\{X_1, \ldots, X_{d+1}\}$ *is strong Rayleigh.*

PROOF: The generating function for the collection $\{X_1, \ldots, X_{d+1}\}$ is just the homogenization of the generating function for $\{X_1, \ldots, X_d\}$. Therefore, what we need to check is that the homogenization of a stable generating function is stable. This is not true for all polynomials[4] but for probability

[4]Indeed $f(x) = x - 1$ is a trivial example of a stable polynomial whose homogenization $f_H(x, y) = x - y$ fails to be stable.

generating functions, or more generally any stable polynomial with nonnegative coefficients, this follows from $(i) \Rightarrow (ii)$ of Proposition 5.3. $\square$

Next we recall the definition of polarization via clone variables from Section 5.3: it is the unique polynomial symmetric separately in each set of clone variables $\{z_{j,1}, \ldots, z_{j,n_j}\}$ such that substituting in z_j for all the clones $z_{j,s}$ yields the original function f. The following result does not rely on f having nonnegative, or even real coefficients.

THEOREM 6.12 (polarization). *The complex polynomial f is stable if and only if its polarization f_p is stable.*

To prove this we require the Grace–Walsh–Szegő theorem. This result may be derived from the theory of stable polynomials, as shown in [**BB09b**] and streamlined in [**Wag11**, Section 4]. We will be content here to quote this century old result; for a proof, see [**COSW04**, Theorem 2.12] or Section 4 of [**Wag11**].

LEMMA 6.13 (GraceWalsh-Szegő Theorem). *Let $f \in \mathbb{C}[z_1, \ldots, z_d]$ be symmetric and multi-affine and let $\mathcal{C}$ be a convex circular region containing the points $\zeta_1, \ldots, \zeta_d$, for instance the upper half-plane $\mathcal{H}$. Then there exists at least one point $\zeta \in \mathcal{C}$ such that $f(\zeta_1, \ldots, \zeta_d) = f(\zeta, \ldots, \zeta)$.* $\square$

PROOF OF THEOREM 6.12: One direction is elementary: if f_p is stable then the diagonalization property allows us to set $z_{j,s} = z_j$ for all j and we deduce stability of f. For the nontrivial direction, suppose f_p is not stable and let $\mathbf{a} := \{a_{j,s}\}$ be numbers in the upper half-plane such that $f_p(\mathbf{a}) = 0$. Fixing the numbers $\{a_{j,s} : j > 1\}$, the Grace-Walsh-Szegő Theorem implies the existence of a number a_1 in the upper half-plane such that $f(a_1, \ldots, a_1, \mathbf{a}') = 0$, where $\mathbf{a}'$ are the numbers $a_{j,s}$ for $j > 1$. Iterating, we arrive at numbers $a_2, \ldots, a_n \in \mathcal{H}$ for which $f(a_1, \ldots, a_n) = 0$, showing that f is not stable. $\square$

A measure μ on $\mathcal{B}_d$ is called **Projected Homogeneous Rayleigh** (PHR) if there is a measure ν on $\mathcal{B}_m$ for some $m \geq d$ such that ν is a homogeneous measure with the Rayleigh property and the projection of ν to the law of the first d variables gives μ. If, furthermore, the law ν is strong Rayleigh, we say that μ is PHSR.

COROLLARY 6.14. *Strong Rayleigh measures are PHSR.*

PROOF: First extend the law μ of the binary variables $\{X_1, \ldots, X_d\}$ by homogenization to a law μ_H on variables $\{X_1, \ldots, X_{d+1}\}$, the last of which takes integer values. By Lemma 6.11 this is still in the class of strong Rayleigh measures. Next replace the integer variable X_{d+1} by clones

$\{X_{d+1,1},\ldots,X_{d+1,d}\}$ by polarization; the resulting measure ν on $\mathcal{B}_{2d}$ is strong Rayleigh by Theorem 6.12 and its projection onto the first d coordinates is μ. □

PROOF OF NEGATIVE ASSOCIATION: To prove negative association, we use an argument due to Feder and Mihail [**FM92**]. This argument shows that homogeneous strong Rayleigh measures (thus by Corollary 6.14, all strong Rayleigh measures) are negatively associated. All that is needed in the Feder–Mihail lemma is for this class to be closed under conditioning on $\{X_j = x\}$ (which is obvious) and have pairwise negative correlations, which we have seen in (5.1). In short, negative association for strong Rayleigh measures is reduced to the following lemma.

LEMMA 6.15 (Feder–Mihail). *Let $\mathcal{S}$ be a class of homogeneous measures on finite Boolean algebras (of differing sizes) which is closed uner conditioning and each of which has pairwise negative correlations. Then all measures in $\mathcal{S}$ are negatively associated.*

Assuming this lemma, the proof of negative association finishes as follows. The class of homogeneous strong Rayleigh measures is closed under conditioning on the value of any variable. We have noted that strong Rayleigh implies negative pairwise correlations, so we conclude from the Feder–Mihail lemma that homogeneous strong Rayleigh measures are negatively associated. Any strong Rayleigh measure is a projection of a homogeneous strong Rayleigh measure, hence inherits the negative association property. It remains only to prove the lemma.

PROOF OF FEDER–MIHAIL LEMMA: Any increasing function on $\mathcal{B}_d$ is a positive linear combination of increasing indicator functions, that is, indicator functions $\mathbf{1}_A$ of upwardly closed events A. It therefore suffices to prove negative correlation for upwardly closed events A and B depending on disjoint sets of variables. (Note: throughout the proof, the phrase "negatively correlated" includes the case of zero correlation.)

We first prove this in the special case where A is the event $\{X_j = 1\}$. We use induction on the number of variables, m. When $m = 2$, the only nontrivial case is $\mathbf{1}_A = X_1$ and $\mathbf{1}_B = X_2$, which are negatively correlated by hypothesis. Now assume for induction that for all measures $\nu \in \mathcal{S}$ on Boolean lattices $\mathcal{B}_d$ for $d \leq m$, for all j, and for all upwardly closed events B not depending on X_j, the functions X_j and $\mathbf{1}_B$ are negatively correlated. Let μ be a law on $\mathcal{B}_{m+1}$, and let $j \leq m+1$ and $B \subseteq \mathcal{B}_{m+1}$ not depending on X_j be given. We will use the inequality

$$pa + (1-p)b \leq p'c + (1-p')d$$

holding when $p, p', a, b, c, d \in [0,1]$ with $p \leq p', a \leq c, b \leq d$ and $a \geq b$. To use this inequality, fix $i \leq m+1$ to be determined later. Let $p := \mu(X_i = 1 \mid X_j = 1)$ and $p' := \mu(X_i = 1 \mid X_j = 0)$, the inequality $p \leq p'$ holds because it is equivalent to pairwise negative correlation of X_i and X_j. We let

$$\begin{aligned} a &:= \mu(B \mid X_i = 1, X_j = 1); \\ b &:= \mu(B \mid X_i = 0, X_j = 1); \\ c &:= \mu(B \mid X_i = 1, X_j = 0); \\ d &:= \mu(B \mid X_i = 0, X_j = 0). \end{aligned}$$

The inequalities $a \leq c$ and $b \leq d$ follow because the conditional measures $(\mu \mid X_j = 1)$, $(\mu \mid X_i = 1)$ and $(\mu \mid X_i = 0)$ are all subject to the induction hypothesis. Finally, to ensure that $a \geq b$, we now choose i judiciously. Equivalent conditions for the inequality $a \geq b$ are

$$\frac{\mu(B, X_i = 1, X_j = 1)}{\mu(B^c, X_i = 1, X_j = 1)} \geq \frac{\mu(B, X_i = 0, X_j = 1)}{\mu(B^c, X_i = 0, X_j = 1)}$$

$$\Updownarrow$$

$$\frac{\mu(B, X_i = 1, X_j = 1)}{\mu(B, X_i = 0, X_j = 1)} \geq \frac{\mu(B^c, X_i = 1, X_j = 1)}{\mu(B^c, X_i = 0, X_j = 1)}$$

$$\Updownarrow$$

$$\mathbb{E}(X_i \mid B, X_j = 1) \geq \mathbb{E}(X_i \mid B^c, X_j = 1).$$

Homogeneity of the measure μ implies that the sum over i of the left and right-hand sides of the last inequality are both equal to the deterministic value $\sum_{i=1}^{m+1} X_i$. Therefore the inequality must hold for at least one $i \leq m+1$. We conclude that

$$\mu(B \mid X_j = 1) = pa + (1-p)b \leq p'c + (1-p')d = \mu(B \mid X_j = 0).$$

Having established that single variables are negatively correlated with upwardly closed events for all measures in $\mathcal{S}$, we consider the general case where A and B are upwardly closed events depending on disjoint sets of variables. Again we use induction, assuming the result for all measures on fewer variables and also already possessing the result in the special case. We run the argument with $\mathbf{1}_A$ in place of X_j, thus $p := \mu(X_i = 1 \mid A)$, $p' := \mu(X_i = 1 \mid A^c)$, and so forth. The inequality $p \leq p'$ is a consequence of the special case. The inequalities $a \leq c$ and $b \leq d$ come from the induction hypothesis applied to the conditional measures $(\mu \mid X_i = 1)$ and $(\mu \mid X_i = 0)$ respectively. Again, constancy of $\sum_{i=1}^{m+1} X_i$ forces the inequality $\mathbb{E}(X_i \mid B, A) \geq \mathbb{E}(X_i \mid B^c, A)$ to hold for some i, finishing the proof. □

6.4.2. *Proof of part (ii): stochastic inequalities.* We begin with the stochastic covering property, which follows directly from Corollary 6.14 and the following lemma, first proved in [**PP11**].

LEMMA 6.16 ([**PP11**, Proposition 2.2]). *If a measure is projected homogeneous strong Rayleigh then it has the stochastic covering property.*

PROOF: Let μ be a projection of a homogeneous strong Rayleigh measure ν. Without loss of generality, we assume that the projection is onto the last d out of m coordinates. Let ν_0 denote the probability measure on $\mathcal{B}_{m-1}$ which is the ν-law of $(X_1, \ldots, X_{m-1})$ conditioned on $X_m = 0$; similarly, let ν_1 denote the probability measure on $\mathcal{B}_{m-1}$ which is the ν-law of $(X_1, \ldots, X_{m-1})$ conditioned on $X_m = 1$. Negative association of ν implies that ν_0 stochastically dominates ν_1: let $B \subseteq \mathcal{B}_m$ be any event not depending on the mth coordinate then B, that is $B = B_* \times \{0, 1\}$ for some $B_* \in \mathcal{B}_{m-1}$; then negative correlation of B and X_m implies that $\nu_0(B_*) \geq \nu_1(B_*)$. The equivalent coupling formulation of stochastic domination (see the beginning of Subsection 6.2.2) shows there is a random variable (X, Y) such that X has law ν_0, Y has law ν_1 and $X \geq Y$. By homogeneity of ν, we see in fact that X always covers Y. Projecting back to $\mathcal{B}_d$, this yields a random pair $(\overline{X}, \overline{Y})$ such that $\overline{X}$ has law $(\mu \,|\, X_m = 0)$, $\overline{Y}$ has law $(\mu \,|\, X_m = 1)$, and $\overline{X}$ is always equal to or covering $\overline{Y}$. □

PROOF OF ULTRA LOG-CONCAVITY OF THE RANK SEQUENCE: Next we examine the rank function. Let μ be a strong Rayleigh measure on $\mathcal{B}_d$. The rank sequence $a_k := \mu(N = k)$ has generating function $g(z) := \sum_{k=0}^{d} a_k z^k = f_\mu(x, \ldots, x)$. This is a stable polynomial because it is a diagonalization of the stable polynomial f_μ, recalling property (d) at the beginning of Section 5. Newton's inequalities (Theorem 4.2) for univariate real stable polynomials then yield ultra log-concavity of the coefficients of the rank sequence. □

PROOF OF RESCALING BY A STABLE COEFFICIENT SEQUENCE: Let $b_0, \ldots, b_d$ be a sequence of nonnegative real numbers whose nonzero elements are an interval $b_r, b_{r+1}, \ldots, b_s$ and are the coefficients of a stable polynoimal. We have seen in Example 4.15 that the sequence $\{b_k\}$ is a multiplier sequence. It follows from Theorem 5.6 that if $f(\mathbf{z}) = \sum_{\mathbf{r}} a_{\mathbf{r}} \mathbf{z}^{\mathbf{r}}$ is a real stable function of $d + 1$ variables then $g(z) := \sum_{\mathbf{r}} a_{\mathbf{r}} b_{r_{d+1}} \mathbf{z}^{\mathbf{r}}$ is also stable.

Now let μ be a strong Rayleigh measure on $\mathcal{B}_d$ with probability generating function f_μ. By Lemma 6.11, the measure ν whose generating function is the homogenization $f_\nu = (f_\mu)_H$ is also stable (recall that ν is the law of $(X_1, \ldots, X_d, d - \sum_{j=1}^{d} X_j)$ when $(X_1, \ldots, X_d)$ has law μ). With $\{b_k : 0 \leq k \leq d\}$ any ultra log-concave sequence as above, we have seen that $g(\mathbf{z}) := \sum_{\mathbf{r}} a_{\mathbf{r}} b_{r_{d+1}} \mathbf{z}^{\mathbf{r}}$ is stable. Setting z_{d+1} equal to 1, we see that $g(\mathbf{z})/g(1, \ldots, 1)$ is a real stable probability generating function, whence the measure described at the end of part *(ii)* of Theorem 6.9 is strong Rayleigh. □

PROOF OF STOCHASTICALLY INCREASING LEVELS: If the interval of support $[r, s]$ of the ultra log-concave sequence $\mathbf{b}$ satisfies $s - r = 0$ or 1, then it is automatically ultra log-concave. Taking $s = r$, we see that if the law μ of $(X_1, \ldots, X_d)$ is strong Rayleigh then so is $(\mu \,|\, N = k)$ as long as $\mu(N = k) > 0$; here again $N := \sum_{j=1}^{d} X_j$. Taking $s = r + 1$ we see that $(\mu \,|\, k \leq N \leq k + 1)$ is strong Rayleigh, provided that the event conditioned on is not null.

Fix k such that $\mu(N=k)$ and $\mu(N=k+1)$ are both nonzero, and let $\mu_k := (\mu \mid N=k)$ and $\mu_{k+1} := (\mu \mid N=k+1)$. The measure $(\mu \mid k \le N \le k+1)$ is strong Rayleigh. Homogenizing by adding the check bit $x_{d+1} := k+1-\sum_{j=1}^d x_j$ gives a measure ν on $\mathcal{B}_{d+1}$ that is also strong Rayleigh. Let $A \subseteq \mathcal{B}_d$ be any upwardly closed set and let $A_* := A \times \{0,1\}$. By negative association, the event A_* is negatively correlated with X_{d+1}. Thus $\nu(A_* \mid X_{d+1}=1) \le \nu(A_* \mid X_{d+1}=0)$, which is equivalent to $\mu_k(A) \le \mu_{k+1}(A)$. Because A was an arbitrary upwardly closed event, this proves that $\mu_k \preceq \mu_{k+1}$. □

6.4.3. *Proof of part (iii): exclusion and other closure properties.* Most of the closure properties in part (*iii*) of Theorem 6.9 are easy, following from the closure properties of the class of stable polynomials listed at the beginning of Section 5. The generating function for the product measure $f_{\mu\times\nu}$ is the product of functions $f_\mu \cdot f_\nu$, whence the strong Rayleigh property for products of strong Rayleigh measures follows from stability of the product of stable polynomials. Projection corresponds to setting some variables equal to 1, whence closure under projections follows from specialization to real values (Lemma 5.2). External fields correspond to replacing f by $cf(a_1z_1,\ldots,a_dz_d)$ for positive real values of the a_j, which follows from the closure of stable polynomials under dilations. To prove closure of the class of strong Rayleigh measures under total symmetrization, there are two avenues. One is to deduce this from the more difficult result for partial symmetrization (see below), observing that total symmetrization may be achieved by repeated partial symmetrization. As pointed out in [**BBL09**, Remark 4.5], a more direct argument is as follows. Let μ be a strong Rayleigh measure on $\mathcal{B}_d$. The sum $N := \sum_{j=1}^d X_j$ has generating function $f_N(z) := f_\mu(z,\ldots,z)$, which is stable because it is a diagonalization of f_μ. The generating function of the total symmetrization $\overline{\mu}$ is the polarization of $f_N(z)$, hence is stable by Theorem 6.12.

The key to proving closure of the class of strong Rayleigh measures under exclusion dynamics is to prove that a single step of **partial symmetrization** preserves the strong Rayleigh property. Let μ be a probability measure on $\mathcal{B}_d$, let $1 \le i < j \le d$ be indices and fix $0 < \theta < 1$. The θ-partial symmetrization of μ with respect to indices i and j is defined to be the measure $\mu_{\theta;i,j}$ whose verbal description is as follows.

> To sample from $\mu_{\theta;i,j}$, first sample from μ, then flip an independent θ-coin to decide whether to transpose the i and j coordinates.

At the level of generating functions, this is Theorem 5.10, which we now recall and prove. The generating polynomial $f_{\mu,\theta;i,j}$ is the one described in Theorem 5.10 by

$$f_{\mu,\theta;i,j} = (1-\theta)f_\mu + \theta\tau(f_\mu)$$

where τ operates by switching arguments i and j. The theorem states that if f is stable then so is $(1-\theta)f + \theta\tau(f)$, which when applied to $f = f_\mu$ yields closure of strong Rayleigh measures under partial symmetrization. We now prove the theorem.

PROOF OF THEOREM 5.10: Let μ be a strong Rayleigh measure. Assume without loss of generality that $i = 1$ and $j = 2$. We need to show that $f_{\mu,\theta,1,2}$ is stable. For every choice of complex numbers $a_3, a_4, \ldots, a_d \in \mathcal{H}$, we know that $f_\mu(\cdot, \cdot, a_3, \ldots, a_d)$ is multi-affine and stable and we need to show that $f_{\mu,\theta,1,2}(\cdot, \cdot, a_3, \ldots, a_d)$ is stable. Therefore it suffices to show:

PROPOSITION 6.17. *For all multi-affine stable functions* $f : \mathbb{C}^2 \to \mathbb{C}$, *the function* $(1-\theta)f(x,y) + \theta f(y,x)$ *is stable.*

REMARK. Because of the restriction to the multi-affine case, we know that $f(x,y) = a + bx + cy + dxy$, so it cannot be too hard to prove this! We observe that if we needed this only for real stable functions, it would follow immediately from the characterization that f is stable if and only if $ad \le bc$, which follows from the mixed partial derivative criterion (5.1). Indeed, the proof of Theorem 5.10 appearing in [**BBL09**, Section 4.4] begins with this and extends via the multivariate Obreschkoff theorem. To keep things self-contained, we will instead derive this from the portion of Theorem 5.8 that we have proved; another self-contained proof is given in [**Lig09**, Theorem 7]; and exhaustive description of this four parameter family is yet another way to finish this.

PROOF OF PROPOSITION 6.17: The operator T on $\mathbb{C}_{\mathrm{ma}}[x,y]$ corresponding to partial symmetrization is defined by its action on the four basis elements $1, x, y$ and xy:

$$T(1) = 1;\; T(x) = (1-\theta)x + \theta y;\; T(y) = (1-\theta)y + \theta x;\; T(xy) = xy\,.$$

Plugging into the definition (5.2) of G_T gives

$$G_T(x,y,u,v) = xy + [(1-\theta)x + \theta y]u + [(1-\theta)y + \theta x]v + uv\,.$$

We need to show this is stable. It is multi-affine and has real coefficients, so we may apply the mixed partial derivative test (5.1). There are six unordered pairs of variables. In each case we easily verify that the left side of (5.1) minus the right side is equal to a nonnegative multiple of a square when $0 \le \theta \le 1$. For example, for the pair (x,u), we evaluate $(\partial^2 f/\partial x\partial u)\,f - (\partial f/\partial x)(\partial f/\partial u)$ to obtain $\theta(y+v)^2$, while for the pair (x,y) we obtain $(\partial^2 f/\partial x\partial y)\,f - (\partial f/\partial x)(\partial f/\partial y) = \theta(1-\theta)(u-v)^2$; up to the symmetries of f, these are the only two cases, therefore the proposition is proved. □

The final step, from partial symmetrization to exclusion dynamics, is a small one. The operators $T_{\theta;i,j}$ defined by $\mu \mapsto \mu_{\theta;i,j}$ generated a semigroup S of operators on the space of probability measures on $\mathcal{B}_d$. For any set $\{\lambda_{\{i,j\}}\}$ of swap rates and any $t \ge 0$, the operator mapping μ to the time t law of the exclusion process with swap rates λ started from the measure μ is in the closure of S. This is more or less self-evident; for a more formal proof, note that when only one swap rate is nonzero, the exclusion evolution operator is already equal to an operator $T_{\theta;i,j}$, after which one can use the Trotter product formula (see [**BBL09**, Proposition 5.1]) to write the general exclusion evolution operator as a limit of products of those with only one nonvanishing swap rate. This completes the proof of Theorem 6.9. □

To finish the discussion of probabilistic applications, we discuss examples. The theorem explicitly addresses exclusion measures (Example 6.5). Conditioned Bernoullis (Example 6.2) are strong Rayleigh due stability of $1-p+pz$ for $0 \leq p \leq 1$, closure under products, (yielding all measures with independent coordinates), and closure under conditioning on $N=k$. Random cluster measures with $q<1$ are conjectured to be Rayleigh and known not to be strong Rayleigh and spanning trees are a special case of determinantal measures, so among the examples 6.2–6.6 it remains only to show that determinantal measures are strong Rayleigh. We will require a result which is the analogue of Example 2.4 but for stability instead of hyperbolicity. The exposition is taken from [**BBL09**, Proposition 3.2] though the result itself has been know for a long time.

PROPOSITION 6.18. *Let $A_1, \dots A_d$ be (complex) positive semi-definite $n \times n$ matrices and let B be a Hermitian matrix, also $n \times n$.*

1. *The polynomial*
$$(z_1, \dots, z_d) \mapsto f(z_1, \dots, z_d) := \text{Det}\,(z_1 A_1 + \cdots + z_d A_d + B)$$
is either identically zero or it is real stable.
2. *If B is also positive semi-definite then the f has all nonnegative coefficients.*
3. *It follows that if $Z := \text{diag}\,(z_1, \dots, z_d)$ and B is any positive semi-definite $d \times d$ matrix, then* Det $(B+Z)$ *is a multi-affine real stable polynomial with all nonnegative coefficients, hence equal to cf_μ for some strong Rayleigh measure μ.*

PROOF: Stability is closed under limits, so it suffices to prove the proposition in the case where the matrices $A_1, \dots, A_d$ are positive definite. The polynomial f is real on real inputs because the determinant of a complex Hermitian matrix is real. Pick a vector $\mathbf{u} \in \mathbb{R}_+^d$ and $\mathbf{v} \in \mathbb{R}^d$. Define a function $\mathbf{z}$ by $\mathbf{z}(t) := \mathbf{v} + t\mathbf{u}$. The matrix $P := \sum_{j=1}^d v_j A_j$ is positive definite, hence invertible with a positive definite square root, call it Q. We may write

$$f(\mathbf{z}(t)) = \text{Det}\,(P)\;\text{Det}\,(tI + QHQ^*)$$

where $H := B + \sum_{j=1}^d u_j A_j$ is Hermitian. Thus $f(\mathbf{z}(t))$ is a polynomial in t that is a constant multiple of the characteristic polynomial of a Hermitian matrix, hence has all real zeros. Because $\mathbf{v} \in \mathbb{R}^d$ and $\mathbf{u} \in \mathbb{R}_d^+$ are arbitrary, the criterion for stability in Proposition 5.1 is satisfied, proving the first statement of the theorem.

Expanding the determinant, we find that the coefficients of f are products of principal minors of the matrices $A_1, \dots, A_d$ and B. When these are all positive semi-definite, the principal minors are positive, proving the second statement. The last statement follows immediately from setting A_i to the matrix with a 1 in the (i,i)-entry and zeros elsewhere, noting that this matrix is positive semi-definite. □

PROOF THAT DETERMINANTAL MEASURES ARE STRONG RAYLEIGH: Assume first that the Hermitian kernel K is invertible. It is easily seen that the generating polynomial f_μ is given by

$$f_\mu(z_1,\ldots,z_d) = \mathrm{Det}\,(I - K + KZ) = \mathrm{Det}\,(K)\,\mathrm{Det}\,(K^{-1} - I + Z)\,,$$

where $Z := \mathrm{diag}\,(z_1,\ldots,z_d)$. Because K is also a contraction, $K^{-1} - I$ is positive semi-definite. The constant $\mathrm{Det}\,(K)$ is positive, so it follows from Example 2.4 that f_μ is stable. The general case may be obtained by taking limits because positive definite kernels are dense in the space of positive semi-definite kernels. □

7. Further applications of stability: determinants, permanents and moments

From the outset, determinants have been prominent in the theory of hyperbolic polynomials. Already in [**Går51**, Example 2], the determinant function on the space of Hermitian matrices was given as an example of a hyperbolic polynomial, with the nonnegative definite matrices being a cone of hyperbolicity (see Example 2.4 above). Related to the determinant but more enigmatic is the **permanent**. The definition of the permanent,

$$\mathrm{Per}\,(A) := \sum_{\sigma\in S_n}\prod_{i=1}^{n} a_{i,\sigma(i)}\,,$$

differs from that of the determinant only in that there is no alternating sign factor $(-1)^{\mathrm{parity}\,(\sigma)}$ in the summand. This makes the permanent much less tractable than the determinant. For example, it is #P-hard to compute the permanent of a zero-one matrix, while determinants may be evaluated in polynomial time. In this section we discuss two results on permanents and one on determinants. The first of these is a lower bound on the permanent, conjectured by van der Waerden in 1926, proved independently by Egorychev and Falikman in 1981, and re-proved in a simpler and more general way by Gurvits in 2009 using the theory of stable functions. The second of these results, the Monotone Column Permanent Conjecture asserts the stability of a certain polynomial obtained as a permanent. It was conjectured in 1999 and proved in 2009. The third result, the so-called BMV conjecture, is only tangentially related to stability theory, but has garnered enough attention to mandate its inclusion here. It was conjectured in 1975. A proof posted recently to the arXiv is believed to be correct.

7.1. The van der Waerden conjecture. One prolific area of research in understanding the permanent has been to identify extremal cases for various families of matrices. Some motivation for understanding permanents comes from graph theory. If A is the incidence matrix of a bipartite graph G, then Per (A) is the number of perfect matchings of G. This interpretation has led to an emphasis on extrema over zero-one matrices. For example, one might consider matrices of zeros and ones with prescribed row sums. An upper bound, conjectured by Minc and proved by Brégman [**Bré73**] is as follows.

THEOREM 7.1 (Brégman's Theorem). *Let A be a nonnegative $n \times n$ matrix of zeros and ones and let $r_1, \ldots, r_n$ denote the row sums of A. Then*

$$\mathrm{Per}\,(A) \leq \prod_{j=1}^{n} (r_j!)^{1/r_j} .$$

□

The lower bound is of course zero because there could be a column of zeros. If we prescribe the column sums as well as the row sums, can we find a nontrivial lower bound? Removing the restriction to zero-one matrices, this question was posed by van der Waerden in 1926 for **doubly stochastic** matrices. A matrix $A = (a_{ij})_{1 \leq i,j \leq n}$ is said to be stochastic if it has nonnegative entries and all row sums $\sum_j a_{ij}$ are equal to 1. The terminology comes from the fact that these matrices are precisely the transition kernels for Markov chains. The matrix A is said to be doubly stochastic if all column sums $\sum_i a_{ij}$ are equal to 1 as well. One might not see at first that the permanent of a doubly stochastic matrix must be nonzero, but this follows from the well known fact that doubly stochastic matrices are positive linear combinations of permutation matrices (and such matrices have permanent equal to 1).

Intuition may suggest that the minimum occurs when all entries are equal to $1/n$. Indeed, van der Waerden conjectured in 1926 that if A is doubly stochastic then

$$\mathrm{Per}\,(A) \geq \frac{n!}{n^n} \tag{7.1}$$

with equality if and only if $a_{ij} = 1/n$ for all i, j. This theorem was proved thirty years ago, independently by Egorychev [**Ego81**] and Falikman [**Fal81**]. Recently, Gurvits [**Gur08**] gave a different proof using stability. Not only does this represent a considerable simplification, but the result is general enough to imply several other well known results. Gurvits begins by identifying the permanent as a coefficient of a polynomial, as follows. If $A = (a_{ij})$ is any $n \times n$ matrix, we may define a homogeneous polynomial p by

$$p(\mathbf{x}) = p_A(\mathbf{x}) := \prod_{i=1}^{n} \sum_{j=1}^{n} a_{ij} x_j .$$

The permanent of A is then the $(1,\ldots,1)$ coefficient of p. Thus,

$$\text{Per}\,(A) = \frac{\partial^n}{\partial x_1 \cdots \partial x_n} p(0,\ldots,0)\,. \tag{7.2}$$

When A is stochastic, each factor in the product evaluates to 1 at $(1,\ldots,1)$. Taking the derivative with respect to x_j and evaluating at $(1,\ldots,1)$ gives the jth column sum, hence if A is doubly stochastic then

$$\frac{\partial}{\partial x_j} p_A(1,\ldots,1) = 1,\ \ 1 \le j \le n\,. \tag{7.3}$$

Let $\mathcal{C}_n$ be the class of homogeneous polynomials of degree n in n variables with all coefficients nonnegative. Following Gurvits, we extend the definition of the term "doubly stochastic" from matrices to $\mathcal{C}_n$ by saying that p is doubly stochastic if (7.3) holds.

To prove the van der Waerden result, we need to derive $\text{Per}\,(A) \ge n!/n^n$ from (7.3). This will not be true for every $p \in \mathcal{C}_n$ satisfying (7.3) but Gurvits' idea was that it should hold for all stable $p \in \mathcal{C}_n$. Because stability is closed under product and each linear polynomial with positive coefficients is stable, we see immediately that p_A is always stable. The other ingredient in Gurvits' proof is to strengthen the induction, replacing the lower bound of $n!/n^n$ by $(n!/n^n)\text{Cap}\,(p)$, where $\text{Cap}\,(p)$ is a constant that evaluates to 1 when $p = p_A$ for a doubly stochastic matrix A.

DEFINITION 7.2. *Let $\mathcal{C}_n$ be the class of homogeneous polynomials of degree n in n variables with all coefficients nonnegative. For $p \in \mathcal{C}_n$, define the capacity of p by*

$$\text{Cap}\,(p) := \inf \frac{p(x_1,\ldots,x_n)}{\prod_{i=1}^n x_i}$$

where the infimum is over nonnegative values of the variables x_i.

PROPOSITION 7.3 ([**Gur08**, Fact 2.2]). *If p is doubly stochastic then* $\text{Cap}\,(p) = 1$.

PROOF: We may view $p \in \mathcal{C}_n$ as a generating function for a probability distribution μ supported on $\Delta_n := \{\omega \in \mathbb{Z}_+^n : |\omega| = n\}$. Letting $\mathbb{E}$ denote expectation with respect to μ we see that (7.3) is equivalent to $\mathbb{E}\omega_j = 1$ for all j.

For $\mathbf{x} \in \mathbb{R}_+^n$, we may evaluate $p(\mathbf{x}) = \log \mathbb{E} \prod_j x_j^{\omega_j}$. Applying Jensen's inequality to the concave function $\log(x)$ gives

$$\begin{aligned}
\log p(x_1,\ldots,x_n) &= \log \mathbb{E} \prod_{j=1}^n x_j^{\omega_j} \\
&\ge \mathbb{E} \log \prod_{j=1}^n x_j^{\omega_j} \\
&= \mathbb{E} \sum_j \omega_j \log x_j\,.
\end{aligned}$$

Because $\mathbb{E}\omega_j = 1$ for all j, this last quantity is just $\log \prod_j x_j$. Thus, on the positive orthant, $p(\mathbf{x}) \geq \prod_{j=1}^n x_j$ showing that $\text{Cap}\,(p) \geq 1$. Setting $x_j = 1$ for all j shows that $\text{Cap}\,(p) \leq 1$. $\square$

Recalling the lower bound on $f'(0)$ for univariate stable polynomials with nonnegative coefficients from Proposition 4.3 it is not hard to envision some kind of induction. The engine of Gurvits' proof, used in the induction step, is the following inequality.

LEMMA 7.4 ([**Gur08**, Theorem 4.10]). *Let $p \in \mathcal{C}_n$ be stable and define*

$$q(x_1, \ldots, x_{n-1}) := \frac{\partial}{\partial x_n} p(x_1, \ldots, x_{n-1}, 0)\,.$$

Then

$$\text{Cap}\,(q) \geq \text{Cap}\,(p) \cdot \left(\frac{d-1}{d}\right)^{d-1} \tag{7.4}$$

where d is the maximum degree of x_n in p.

PROOF: Letting $x_1, \ldots, x_{n-1}$ range over positive numbers whose product is 1, we need to show that

$$q(x_1, \ldots, x_{n-1}) \geq \text{Cap}\,(p) G(d)$$

where $d \leq n$ is the maximum degree of x_n in p and $G(m) := [(m-1)/m]^{m-1}$. Fix $x_1, \ldots, x_{n-1}$. The specialization property implies that the univariate polynomial $R(t) := p(x_1, \ldots, x_{n-1}, t)$ is stable. By definition of capacity, $\text{Cap}\,(p) \leq \inf_{t>0} R(t)/t$. The degree of R is equal to d, whence it follows from Proposition 4.3 that

$$q(x_1, \ldots, x_{n-1}) = R'(0) \geq \text{Cap}\,(p) G(d)\,.$$

$\square$

We may now give Gurvits' result implying the van der Waerden conjecture.

THEOREM 7.5. *Let p be stable with nonnegative coefficients. Then*

$$\frac{\partial^n}{\partial x_1 \cdots \partial x_n} p(0, \ldots, 0) \geq \text{Cap}\,(p) \frac{n!}{n^n}\,.$$

PROOF: Let $q_n = p$ and in general define

$$q_i(x_1, \ldots, x_i) := \frac{\partial^{n-i}}{\partial x_{i+1} \cdots \partial x_n} p(x_1, \ldots, x_i, 0 \ldots, 0)\,.$$

Stability is closed under differentiation and specializing to real values, hence by induction on $n - i$, each q_i is stable. Also, $q_i \in \mathcal{C}_i$ for each i. Applying Lemma 7.4 with q_i in place of p and i in place of n shows that

$$\text{Cap}\,(q_{i+1}) \geq \text{Cap}\,(q_i) G(i)$$

because i is an upper bound for the degree of x_i in q_i. Inductively, we see that

$$\text{Cap}\,(q_1) \geq \text{Cap}\,(p) \prod_{i=2}^{n} G(i)\,. \tag{7.5}$$

The right-hand side may be identified as

$$\prod_{i=2}^{n} G(i) = n!/n^n\,. \tag{7.6}$$

To identify the left-hand side, observe that $q_1(x_1)$ is homogeneous of degree 1 in x_1, that is, $q_1 = cx_1$. Thus

$$\text{Cap}\,(q_1) = c = q_1'(0) = \frac{\partial^n}{\partial x_1 \cdots \partial x_n} p(0, \ldots, 0)\,. \tag{7.7}$$

Plugging (7.7) and (7.6) into (7.5) proves the theorem. □

Another result that succumbs to this method is the Schrijver–Valiant conjecture, proved in 1998. This concerns an integer version of doubly stochastic matrices. Let $\Lambda(k, n)$ denote the collection of $n \times n$ matrices whose entries are nonnegative integers and whose rows and columns all sum to k. Define

$$\lambda(k, n) = k^{-n} \min\{\text{Per}\,(A) : A \in \Lambda(k, n)\}$$

and

$$\theta(k) = \lim_{n \to \infty} \lambda(k, n)^{1/n}\,.$$

In 1980 it was proved [**SV80**] that $\theta(k) \leq G(k)$ and equality was conjectured. The proof by Schrijver [**Sch98**] was difficult. This result is a corollary of Theorem 7.5. For details on this, as well as a separate application of Theorem 7.5 to prove a lower bound on the mixed discriminant, see [**Gur08**].

7.2. The Monotone Column Permanent Conjecture. Say that the $n \times n$ matrix A is a **monotone column matrix** if its entries are real and weakly decreasing down each column, that is, $a_{i,j} \geq a_{i+1,j}$ for $1 \leq i \leq n - 1$ and $1 \leq j \leq n$. Let J_n denote the $n \times n$ matrix of all ones. It was conjectured in [**HOW99**] that whenever A is a monotone column matrix, the univariate polynomial $\text{Per}\,(zJ_n + A)$ has only real roots. A proof was given there for the case where A is a zero-one matrix.

Although there was strong intuition already present, the proof had to wait for the development of multivariate stable function theory. In 2009, Brändén, Haglund, Visontai and Wagner [**BHVW09**] were able to prove this conjecture by proving something stronger, namely multivariate stability.

THEOREM 7.6 (Multivariate Monotone Column Permanent Theorem (MMCPT)). *Let Z_n be the diagonal matrix whose entries are the n indeterminates $z_1, \ldots, z_n$. Let A by an $n \times n$ monotone column matrix. Then*

Per $(J_n Z_n + A)$ *is a stable polynomial in the variables* $z_1, \ldots, z_n$*. Specializing to* $z_j \equiv z$ *for all* j *preserves stability, hence the original conjecture follows.*

I will give only a brief sketch of the proof, which relies on two results proved by Borcea and Brändén. One of these is the criterion, Theorem 5.8 for a linear operator on multi-affine polynomials to be stability preserving. They use the sufficiency direction, namely that stability of the $(2n)$-variable "algebraic symbol" $G_T(\mathbf{z}, \mathbf{w})$ implies that T is a stability preserving operator on $\mathbb{C}_{\text{ma}}[z_1, \ldots, z_d]$; this is the direction for which a proof was included above. The other lemma they require is the following result.

LEMMA 7.7 ([**BHVW09**, Proposition 2.5]). *Let* V *be a real vector space and let* $\phi : V^n \to \mathbb{R}$ *be a multilinear form. Let* $e_1, \ldots, e_n, v_2, \ldots, v_n$ *be fixed vectors in* V *and suppose that the polynomial*

$$\phi(e_1, v_2 + z_2 e_2, \ldots, v_n + z_n e_n)$$

is not identically zero in $\mathbb{R}[z_2, \ldots, z_n]$*. Then the set of all* $v_1 \in V$ *for which the polynomial*

$$\phi(v_1 + z_1 e_1, \ldots, v_n + z_n e_n)$$

is stable is either empty or a convex cone over the origin containing $\pm e_1$*.* □

By means of this lemma, the MMCPT may be reduced to the special case where the entries of A are all zero or one. I will not reproduce this reduction here, but the main idea is an induction. If we have proved the result in the case where k columns are real monotone and $n - k$ are zero-one monotone, then applying the lemma to one of the $n - k$ columns shows that stability is achieved over a convex cone containing all zero-one monotone columns, and such a cone necessarily contains all real monotone columns.

After a change of variables to $y_j = 1 + 1/z_j$ and the introduction of new variables $x_1, \ldots, x_n$, stability of Per $(z_j + a_{ij})$ will follow if we show stability of Per $(a_{ij} y_j + (1 - a_{ij} x_i)$ for monotone zero-one matrices. Denote this last permanent as Per $(B(A))$. Expanding the permanent on the last row gives a differential recurrence relation which may be written in form $f_j = T f_{j-1}$ where f_j is the permanent of the upper $j \times j$ submatrix of $B(A)$. Here T is an operator of the form $k + z_j \sum_{i=1}^{j-1} \partial / \partial z_j$. The sufficiency criterion reduces the task to checking stability of the algebraic symbol, G_T, which may be accomplished by a simple computation.

7.3. The BMV conjecture. In 1975, Bessis, Moussa and Villani [**BMV75**] formulated what is now know as the BMV conjecture.

CONJECTURE 7.8 (BMV). *Let* A *and* B *be Hermitian* $n \times n$ *matrices with* B *positive semi definite. Then the function*

$$\lambda \mapsto \text{Tr} \exp (A - \lambda B)$$

is the Laplace transform of a positive measure on $[0, \infty)$*; here,* Tr $(\cdot)$ *denotes the trace.*

This seems vaguely tied to a number of the themes we have been discussing, but not necessarily related in a direct way to the theory of stable functions. In fact, both the conclusion and the hypotheses will reveal more upon further scrutiny. Beginning with the obvious, we recall that the linear cone of positive semi-definite matrices is a cone of hyperbolicity for the determinant function on the space of Hermitian matrices. Thus if we take the determinant instead of the trace of the exponential, the resulting function $\lambda \mapsto \text{Det}\,(A - \lambda B)$ is stable.

Recall from Edrei's Equivalence Theorem (Theorem 4.9) that stability of the polynomial $f := \sum_{j=0}^{k} a_j z^j$ is equivalent to a sequence of inequalities, which can be summed up by saying that $\{a_k\}$ is a Pólya frequency sequence. Similarly, a function ϕ on $[0, \infty)$ is a Laplace transform of a positive measure on $\mathbb{R}^+$ if and only ϕ is a **completely monotone** function, meaning that the derivatives of ϕ do not change sign on $[0, \infty)$ and alternate in sign: for each nonnegative integer n and positive real t,

$$(-1)^n \frac{d^n}{dt^n} \phi(t) \geq 0 \,.$$

This result is the Bernstein–Widder Theorem [**Ber28**]. Evidently the conclusion is a property whose definition has features in common with stability, a property we know to be true for $\lambda \mapsto \text{Det}\,(A - \lambda B)$.

In 2004, Lieb and Seiringer [**LS04**] found an equivalent formulation of the BMV conjecture that looks even more similar to stability theory.

THEOREM 7.9. *The BMV conjecture is equivalent to the polynomials*

$$\text{Tr}\,(A + \lambda B)^n$$

having nonnegative coefficients for all integers $n \geq 1$ and all pairs of matrices A, B that are both *positive semi-definite.*

The exponential is gone, the coefficient on λ is positive, only integer powers are involved, and what is needed is a countable sequence of inequalities: all coefficients in a sequence of polynomials must be nonnegative. Another equivalence brings the BMV conjecture directly into the realm of stable function theory. Let $p(x, y)$ be any stable bivariate polynomial with nonnegative coefficients and nonzero constant term. Let a_{mn} be the Taylor coefficients of $(\partial p/\partial x)/p$. The BMV conjecture is equivalent to the proposition that $(-1)^{m+n} a_{mn} \geq 0$ for all m, n.

In July, 2011, a proof of the BMV conjecture by H. Stahl [**Sta11**] was posted on arXiv. At the time of writing, the refereed publication has not appeared, but word on the street is that the proof should be correct. The proof is sufficiently complicated that understanding its specialization to the 3×3 case is considered a good project, perhaps for a dissertation. There is a sense that there may be a simpler proof out there, and that the "book proof" of the BMV-Stahl Theorem is likely to involve stable polynomials.

Acknowledgements

I owe a huge debt to Yuliy Baryshnikov for helping me to understand the material in Part I and to Petter Brändén for helping me to understand the material in Part II. Without their help, this survey could not have been written. Thanks are also due to Mirkó Visontai for his careful reading and comments, to Alan Sokal for crucial help and corrections on the development of multivariate stability, and to David Wagner for helpful discussions.

References

[ABG70] M. Atiyah, R. Bott, and L. Gårding. Lacunas for hyperbolic differential operators with constant coefficients, I. *Acta Mathematica*, 124:109–189, 1970.

[AS08] N. Alon and J. Spencer. *The probabilistic method*. John Wiley & Sons, Hoboken, New Jersey, third edition, 2008.

[BB09a] J. Borcea and P. Brändén. The Lee-Yang and Pólya-Schur programs, I: Linear operators preserving stability. *Invent. Math.*, 177:541–569, 2009.

[BB09b] J. Borcea and P. Brändén. The Lee-Yang and Pólya-Schur programs, II: Theory of stable polynomials and applications. *Comm. Pure Appl. Math.*, 62:1595–1631, 2009.

[BB10] J. Borcea and P. Brändén. Multivariate Pólya-Schur classification problems in the Weyl algebra. *Proc. London Math. Soc.*, 101:73–104, 2010.

[BBBP08] Y. Baryshnikov, W. Brady, A. Bressler, and R. Pemantle. Two-dimensional quantum random walk. *arXiv*, http://front.math.ucdavis.edu/0810.5495: 34 pages, 2008.

[BBL09] J. Borcea, P. Brändén, and T. Liggett. Negative dependence and the geometry of polynomials. *J. AMS*, 22:521–567, 2009.

[BC92] J. Brown and C. Colbourn. Roots of the reliability polynomial. *SIAM J. Disc. Math.*, 5:571–585, 1992.

[Ber28] S. Bernstein. Sur les fonctions absolument monotones. *Acta Math.*, 52:1–66, 1928.

[BHVW09] P. Brändén, J. Haglund, M. Visontai, and D. Wagner. Proof of the monotone column permanent conjecture. In *FPSAC 2009*, volume AK, pages 443–454, Nancy, 2009. Assoc. Discrete Math. Theor. Comput. Sci.

[BMV75] D. Bessis, P. Moussa, and M. Villani. Monotonic converging variational approximations to the functional integrals in quantum statistical mechanics. *J. Math. Phys.*, 16:2318–2325, 1975.

[BP11] Y. Baryshnikov and R. Pemantle. Asymptotics of multivariate sequences, part iii: quadratic points. *Adv. Math.*, 228:3127–3206, 2011.

[Brä07] P. Brändén. Polynomials with the half-plane property and matroid theory. *Adv. Math.*, 216:302–320, 2007.

[Brä11] P. Brändén. Iterated sequences and the geometry of zeros. *J. Reine Angew. Math.*, 658:115–131, 2011.

[Bré73] L. Brégman. Some properties of nonnegative matrices and their permanents. *Sov. Math. Dokl.*, 14:945–949, 1973.

[Bre89] F. Brenti. Unimodal, log-concave and Pólya frequency sequences in combinatorics. *Memoirs of the AMS*, 413:106+viii, 1989.

[Bre94] F. Brenti. Log-concanve and unimodal sequences in algebra, combinatorics and geometry: an update. In *Jerusalem Combinatorics '93*, volume 178, pages 71–89, Providence, 1994. AMS.

[Bre95] F. Brenti. Combinatorics and total positivity. *J. Comb. Theory, ser. A*, 71:175–218, 1995.

[CC95] T. Craven and G. Csordas. Complex zero decreasing sequences. *Meth. Appl. Anal.*, 2:420–441, 1995.

[Con78] John B. Conway. *Functions of One Complex Variable*. Springer-Verlag, New York, second edition, 1978.

[COSW04] Y.-B. Choe, J. Oxley, A. Sokal, and D Wagner. Homogeneous multivariate polynomials with the half-plane property. *Adv. Appl. Math.*, 32:88–187, 2004.

[Cso03] G. Csordas. Complex zero decreasing sequences and the Riemann hypothesis, II. In *ISSAC 2001*, volume 10, pages 121–134, Dordrecht, 2003. Kluwer Academic Publishers.

[CV90] G. Csordas and R. Varga. Necessary and sufficient conditions and the Riemann hypothesis. *Adv. Appl. Math*, 11:328–357, 1990.

[Edr53] A. Edrei. Proof of a conjecture on Schoenberg on the generating function of a totally positive sequence. *Canadian J. Math.*, 5:86–94, 1953.

[Ego81] G. Egorychev. The solution of van der Waerden's problem for permanents. *Adv. Math.*, 42:299–305, 1981.

[Fal81] D. Falikman. Proof of the van der Waerden's conjecture on the permanent of a doubly stochastic matrix. *Mat. Zametki*, 29:931–938, 1981. (in Russian).

[FB85] A. Fettweis and S. Basu. On discrete scattering Hurwitz polynoimals. *International J. Circuit Theory Appl.*, 13:47–59, 1985.

[FKG71] C. Fortuin, P. Kasteleyn, and P. Ginibre. Correlation inequalities on some partially ordered sets. *Comm. Math. Phys.*, 22:89–103, 1971.

[FM92] T. Feder and M. Mihail. Balanced matroids. *Annual ACM Symposium on Theory of Computing*, pages 26–38, 1992.

[Går51] L. Gårding. Linear hyperbolic partial differential equations with constant coefficients. *Acta Math.*, 85:1–62, 1951.

[Går59] L. Gårding. An inequality for hyperbolic polynomials. *J. Math. Mech.*, 8:957–965, 1959.

[GJ83] I. P. Goulden and D. M. Jackson. *Combinatorial enumeration*. A Wiley-Interscience Publication. John Wiley & Sons Inc., New York, 1983. With a foreword by Gian-Carlo Rota, Wiley-Interscience Series in Discrete Mathematics.

[GK71] C. Gruber and H. Kunz. General properties of polymer systems. *Comm. Math. Phys.*, 22:133–161, 1971.

[GKZ94] I. Gelfand, M. Kapranov, and A. Zelevinsky. *Discriminants, Resultants and Multidimensional Determinants*. Birkhäuser, Boston-Basel-Berlin, 1994.

[Gül97] O. Güler. Hyperbolic polynomials and interior point methods for convex programming. *Math. Oper. Res.*, 22:350–377, 1997.

[Gur06] L. Gurvits. The van der Waerden conjecture for mixed discriminants. *Adv. Math.*, 200:435–454, 2006.

[Gur08] L. Gurvits. Van der Waerden/Schrijver-Valiant like conjectures and stable (aka hyperbolic) homogeneous polynomials: one theorem for all. *Elec. J. Combin.*, 15:26, 2008.

[Hen91] Peter Henrici. *Applied and Computational Complex Analysis. Vol. 2*. John Wiley & Sons Inc., New York, 1991. Special functions—integral transforms—asymptotics—continued fractions, Reprint of the 1977 original, A Wiley-Interscience Publication.

[HKPV09] J. B. Hough, M. Krishnapur, Y. Peres, and B Virag. *Zeros of Gaussian analytic functions and determinantal point processes*, volume 51 of *University Lecture Series*. American Mathematical Society, Providence, RI, 2009.

[HL72] O. Heilmann and E. Lieb. Theory of monomer-dimer systems. *Comm. Math. Phys.*, 25:190–232, 1972.

[Hör83] Lars Hörmander. *The analysis of linear partial differential operators. I.* Springer-Verlag, Berlin, 1983. Distribution theory and Fourier analysis.

[HOW99] J. Haglund, K. Ono, and D. G. Wagner. Theorems and conjectures involving rook polynomials with only real zeros. In *Topics in Number Theory*, volume 467 of *Math. Appl.*, pages 207–221. Kluwer Academic Publishers, Dordrecht, 1999.

[Hur96] A. Hurwitz. On the conditions under which an equation has only roots with negative real parts. In *Stability Theory*, volume 121 of *International Series of Numerical Mathematics*, pages 239–249, Berlin, 1996. Birkhäuser. Reprinted from *Math. Ann.* **46** (1895).

[Kar68] S. Karlin. *Total Positivity, vol. 1.* Stanford Univeristy Press, Stanford, CA, 1968.

[Lig85] T. Liggett. *Interacting Particle Systems*, volume 276 of *Grundlehren der mathematischen Wissenschaften.* Springer-Verlag, New York, 1985.

[Lig09] T. Liggett. Distributional limits for the symmetric exclusion process. *Stoch. Proc. Appl.*, 119:1–15, 2009.

[LS81] E. Lieb and A. Sokal. A general Lee-Yang theorem for one-component and multicomponent ferromagnets. *Comm. Math. Phys.*, 80:153–179, 1981.

[LS04] E. Lieb and R. Seiringer. Equivalent forms of the Bessis-Moussa-Villani conjecture. *J. Stat. Phys.*, 115:185–190, 2004.

[LY52] T.D. Lee and C.N. Yang. Statistical theory of equations of state and phase transitions, II: lattice gas and Ising model. *Phys. Rev.*, 87:410–419, 1952.

[Lyo03] R. Lyons. Determinantal probability measures. *IHES*, 98:167–212, 2003.

[MS10] P. McNamara and B. Sagan. Infinite log-concavity: developments and conjectures. *Adv. Appl. Math.*, 44:1–15, 2010.

[Nij76] N. Nijenhuis. On permanents and the zeros of rook polynomials. *J. Comb. Theory, ser. A*, 21:240–244, 1976.

[NN93] Yu. Nesterov and A. Nemirovskii. *Interior point polynomial methods in convex programming*, volume 13 of *SIAM Studies in Applied Mathematics.* SIAM Publications, Philadlephia, PA, 1993.

[Pem91] R. Pemantle. Choosing a spanning tree for the integer lattice uniformly. *Ann. Probab.*, 19:1559–1574, 1991.

[Pem00] R. Pemantle. Toward a theory of negative dependence. *J. Math. Phys.*, 41:1371–1390, 2000.

[Pet45] I. G. Petrovsky. On the diffusion of waves and the lacunas for hyperbolic equations. *Mat. Sb.*, 17:289–370, 1945.

[PP11] R. Pemantle and Y. Peres. Concentration of Lipschitz functionals of determinantal and other strong Rayleigh measures. *arxiv*, 2011.

[PS14] G. Pólya and J. Schur. Über zwei Arten von Faktorenfolgen in der Theorie der algebraischen Gleichungen. *J. Reine Angew. Math.*, 144:89–113, 1914.

[PS97] A. Panconesi and A. Srinivasan. Randomized distributed edge coloring via an extension of the Chernoff–Hoeffding bounds. *SIAM J. Comput.*, 26:350–368, 1997.

[PW02] R. Pemantle and M.C. Wilson. Asymptotics of multivariate sequences. I. Smooth points of the singular variety. *J. Combin. Theory Ser. A*, 97(1):129–161, 2002.

[PW04] R. Pemantle and M.C. Wilson. Asymptotics of multivariate sequences, II. Multiple points of the singular variety. *Combin. Probab. Comput.*, 13:735–761, 2004.

[PW08] R. Pemantle and M.C. Wilson. Twenty combinatorial examples of asymptotics derived from multivariate generating functions. *SIAM Review*, 50:199–272, 2008.

[RS01] G. Royle and A. Sokal. The Brown-Colbourn conjecture on zeros of reliability polynomials is false. *J. Comb. Theory, ser. B*, 94:345–360, 2001.

[RW08] Alexander Raichev and Mark C. Wilson. A new approach to asymptotics of Maclaurin coefficients of algebraic functions. Report CDMTCS-322, Centre for Discrete Mathematics and Theoretical Computer Science, University of Auckland, New Zealand, April 2008. http://www.cs.auckland.ac.nz/CDMTCS/researchreports/322alexmcw.pdf.

[Saf00] K. Safonov. On power series of algebraic and rational functions. *J. Math. Anal. Appl.*, 243(2):261–277, 2000.

[Sch47] I. Schoenberg. On totally positive functions, Laplace integrals, and entir e functions of the Lagurre–Pólya–Schur type. *Proc. Nat. Acad. Sci. USA*, 33:11–17, 1947.

[Sch98] A. Schrijver. Counting 1-factors in regular bipartite graphs. *J. Comb. Theory, ser. B*, 72:122–135, 1998.

[Sok01] A. Sokal. Bounds on the complex zeros of (di)chromatic polynomials and Potts model partition functions. *Comb. Prob. Comput.*, 10:41–77, 2001.

[Sok05] A. Sokal. The multivariate tutte polynomial (alias potts model) for graphs and matroids. In *Surveys in Combinatorics*, volume 327 of *London Math. Soc. Lecture Note Series*, pages 173–226. Cambridge University Press, Cambridge, 2005.

[Sok11] A. Sokal. Roots of a formal power series, with applications to graph enumeration an q-series, lectures at Queen Mary University of London. http://www.maths.qmul.ac.uk/~pjc/csgnotes/sokal/, March-April 2011.

[Sta89] R. Stanley. Log-concanve and unimodal sequences in algebra, combinatorics and geometry. *Annals of the New York Academy of Sciences*, 576:500–535, 1989.

[Sta11] H. Stahl. Proof of the BMV conjecture. *arXiv*, http://front.math.ucdavis.edu/1107.4875: 56 pages, 2011.

[SV80] A. Schrijver and W. Valiant. On lower bounds for permanents. *Indag. Math.*, 42:425–427, 1980.

[Sve68] L. Svensson. Necessary and sufficient conditions for the hyperbolicity ofpolynomials with hyperbolic principal part. *Ark. Mat.*, 8:145–162, 1968.

[Wag00] D. G. Wagner. Zeros of reliability polynomials and f-vectors of matroids. *Comb. Prob. Comput.*, 9:167–190, 2000.

[Wag06] D. G. Wagner. Matroid inequalities from electrical network theory. *Elec. J. Combin.*, 11:17, 2004/2006.

[Wag08] D. G. Wagner. Negatively correlated random variables and mason's conjecture for independent sets in matroids. *Ann. Comb.*, 12:211–239, 2008.

[Wag11] D. G. Wagner. Multivariate stable polynomials: theory and application. *Bull. Amer. Math. Soc.*, 48:53–84, 2011.

[Wid41] D. Widder. The laplace transform. page 406, 1941.

[Wu88] F. Wu. Potts model and graph theory. *J. Stat. Phys.*, 52:99–112, 1988.

[WW09] D. Wagner and Y. Wei. A criterion for the half-plane property. *Disc. Math.*, 309:1385–1390, 2009.

[YL52] C.N. Yang and T.D. Lee. Statistical theory of equations of state and phase transitions, I: Theory of condensation. *Phys. Rev.*, 87:404–409, 1952.

Current Developments in Mathematics, 2011

Introduction to KPZ

Jeremy Quastel

ABSTRACT. This is an introductory survey of the Kardar-Parisi-Zhang equation (KPZ). The first chapter provides a non-rigorous background to the equation and to some of the many models which are supposed to lie in its universality class, as well as the predicted, non-standard fluctuations. The second chapter provides a rigorous introduction to the stochastic heat equation, whose logarithm is the solution of KPZ, as well as some of the known methods for proving convergence of discrete growth models and directed polymer free energies. Finally, we end with a sketch of the derivation of exact formulas for the one-point distributions of KPZ at finite time for special initial data, from the Tracy-Widom formulas for asymmetric exclusion.

CONTENTS

1991 *Mathematics Subject Classification.* 60B20, 60F05, 60H15, 60K35, 82C22, 82C44.

1. A physical introduction

1.1. KPZ/Stochastic Burgers/Scaling exponent. The Kardar-Parisi-Zhang (KPZ) equation is,

$$\partial_t h = -\lambda(\partial_x h)^2 + \nu\partial_x^2 h + \sqrt{D}\xi \tag{1}$$

where ξ denotes space-time white noise which is the distribution valued Gaussian field with correlation function

$$\langle \xi(t,x), \xi(s,y)\rangle = \delta(t-s)\delta(x-y). \tag{2}$$

It is an equation for a randomly evolving height function $h \in \mathbb{R}$ which depends on position $x \in \mathbb{R}$ and time $t \in \mathbb{R}_+$. λ, ν and D are physical constants.

The equation was introduced by Kardar, Parisi and Zhang in 1986 [**80**], and quickly became the default model for random interface growth in physics. Mathematically, the non-linearity was too large to handle by existing methods of stochastic partial differential equations. So there is a very serious problem of well-posedness.

Formally, it is equivalent to the stochastic Burgers equation

$$\partial_t u = -\lambda\partial_x u^2 + \nu\partial_x^2 u + \sqrt{D}\partial_x\xi. \tag{3}$$

which, if things were nice, would be satisfied by $u = \partial_x h$.

An analogous equation can be written in higher dimensions,

$$\partial_t h = -\lambda |\nabla h|^2 + \nu \Delta h + \sqrt{D}\xi \tag{4}$$

with $x \in \mathbb{R}^d$. One can also attempt to generalize the non-linearity

$$\partial_t h = F(\nabla h) + \nu \Delta h + \sqrt{D}\xi \tag{5}$$

It appears that with space-time white noise forcing, *only* (1) yields a non-trivial field. We will restrict ourselves here to the one space dimension with the quadratic non-linearity (1). Even in this $1+1$ dimensional situation we are still in the very difficult case of a field theory with broken time reversible invariance.

The stochastic Burgers equation (3) is a toy model for turbulence. A dynamical renormalization group analysis was performed on it in 1977, by Forster, Nelson and Stephen [**60**] (see also [**80**], [**118**]), predicting a dynamical scaling exponent

$$z = 3/2. \tag{6}$$

For the solution h of the KPZ equation this means that one expects non-trivial fluctuation behaviour under the rescaling

$$h_\epsilon(t, x) = \epsilon^{1/2} h(\epsilon^{-z} t, \epsilon^{-1} x). \tag{7}$$

We will discuss this much more precisely later in the notes.

Now we move to the physical derivation of the process and the physical predictions. Note that this entire introduction is not intended to be rigorous, but just to sketch the physical background. From a mathematical point of view, one is of course interested in proving existence and uniqueness for the equation. On the other hand, we will see that the solutions can be written in terms of a classically well-posed stochastic partial differential equation. So the main issue for these notes will be on the actual behaviour of solutions, the scaling exponents, and the large scale fluctuation behaviour which is conjectured to be universal within what is called the KPZ universality class. For this reason we will spend in Section 1 a great deal of time discussing the physical picture. The mathematics begins in Section 2.

1.2. Physical derivation. h grows by random deposition as well as diffusion. The change in time has three contributions: (1) Slope dependent, or lateral growth, (2) Relaxation, (3) Random forcing. The equation then reads

$$\partial_t h = -\lambda F(\partial_x h) + \nu \partial_x^2 h + \sqrt{D}\xi \tag{8}$$

The $\partial_x^2 h$ term (2) represents the simplest possible form of relaxation/smoothing/diffusion. ν is the diffusivity, or viscosity. The random forcing (3) is assumed to be roughly independent at different positions and different times. The simplest model is Gaussian space-time white noise, which has mean zero and space-time correlations

$$\langle \xi(t,x), \xi(s,y) \rangle := E[\xi(t,x)\xi(s,y)] = \delta(t-s)\delta(x-y). \tag{9}$$

$\sqrt{D}$ represents the strength of the noise. Traditionally it has a square root so that D is the mean square, or variance. The key term (1), the deterministic part of the growth, is assumed to be a function only of the slope, and to be a symmetric function. Here is a picture of what we mean by lateral growth

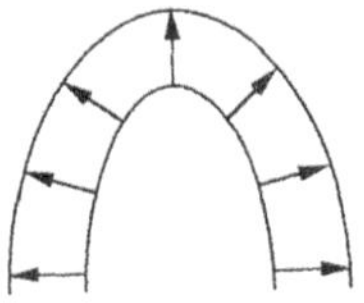

From the picture, the natural choice for F might be $(1+|\partial_x h|^2)^{-1/2}$, however this leads to a seemingly intractable equation. In fact what is done is to take a general F and expand

$$F(s) = F(0) + F'(0)s + \tfrac{1}{2}F''(0)s^2 + \cdots \tag{10}$$

The first term can be removed from the equation by a time shift. The second should vanish because of the symmetry, but can anyway be removed from the equation by a constant velocity shift of coordinates. Thus the quadratic term is the first nontrivial contribution, and it is the only one kept. We arrive at the KPZ equation,

$$\partial_t h = -\lambda(\partial_x h)^2 + \nu\partial_x^2 h + \sqrt{D}\xi. \tag{11}$$

There is something wrong with this derivation. The problem is that $|\partial_x h|^2$ is not small. As we will see in 1.6, it is huge. So one needs to subtract a huge term reflecting the small scale fluctuations. Amazingly, through such a naive derivation, one finds a non-trivial field (see [**84**], [**69**], [**12**] for introductions.)

1.3. Scaling. We can restrict attention to the special choice $\lambda = \nu = \frac{1}{2}$, $D = 1$, because if h satisfies

$$\partial_t h = -\tfrac{1}{2}(\partial_x h)^2 + \tfrac{1}{2}\partial_x^2 h + \xi \tag{12}$$

then we can write

$$h_\epsilon(t,x) = \epsilon^b h(\epsilon^{-z}t, \epsilon^{-1}x) \tag{13}$$

and we have $\partial_t h = \epsilon^{z-b}\partial_t h_\epsilon$, $\partial_x h = \epsilon^{1-b}\partial_x h_\epsilon$ and $\partial_x^2 h = \epsilon^{2-b}\partial_x^2 h_\epsilon$. More interesting is how the white noise rescales,

$$\xi(t,x) \overset{\text{dist}}{=} \epsilon^{\frac{z+1}{2}}\xi(\epsilon^{z}t, \epsilon^{1}x), \tag{14}$$

where the equality is in the sense that the two random fields have the same distribution. This leads to

$$\partial_t h_\epsilon = -\tfrac{1}{2}\epsilon^{2-z-b}(\partial_x h_\epsilon)^2 + \tfrac{1}{2}\epsilon^{2-z}\partial_x^2 h_\epsilon + \epsilon^{b-\frac{1}{2}z+\frac{1}{2}}\xi. \tag{15}$$

Clearly we can now choose $\lambda = \frac{1}{2}\epsilon^{2-z-b}$, $\nu = \frac{1}{2}\epsilon^{2-z}$, $\sqrt{D} = \epsilon^{b-\frac{1}{2}z+\frac{1}{2}}$ to get (1) from (12). When comparing discrete models to KPZ, one has to identify the appropriate λ, ν and D (see [**111**] for a discussion.)

1.4. Formal invariance of Brownian motion. Linearizing (11) one obtains the Langevin equation,

$$\partial_t h = \nu \partial_x^2 h + \sqrt{D}\xi \tag{16}$$

whose solution is the infinite dimensional Ornstein-Uhlenbeck process,

$$h(t,x) = \int_{\mathbb{R}} p_\nu(t, x-y)h(0,y)dy + \sqrt{D}\int_0^t \int_{\mathbb{R}} p_\nu(t-s, x-y)\xi(s,y)dyds \tag{17}$$

where

$$p_\nu(t,x) = \frac{1}{\sqrt{4\pi\nu t}}e^{-x^2/4\nu t}. \tag{18}$$

Two sided Brownian motion $B(x)$, $x \in \mathbb{R}$ normalized to have

$$E[(B(y) - B(x))^2] = (2\nu)^{-1}D|y-x| \tag{19}$$

can be checked to be invariant for this process.

Remarkably, the two-sided Brownian motion is almost invariant for (11) as well. The only sticky point is that the two sided Brownian motion $B(x)$ will have a global height shift as time proceeds. In particular, what is true is that the probability measure corresponding to the distributional derivative of the Brownian motion $B'(x)$ (which is another white noise) is invariant for the stochastic Burgers equation (3) [**15**]. Or at the level of KPZ, the measure corresponding to $B(x)+N$, where N is given by Lebesgue measure, is invariant for (11). Note, however, that the latter is not a probability measure (it is "non-normalizable"). What we mean is that the product measure of white noise and Lebesgue measure on $\mathbb{R}$ is invariant for the process $(h'(x), h(0))$ [**63**]. Remember that when we say Brownian motion is almost invariant, we mean that if $h_0(x)$ is a two sided Brownian motion, then, for fixed t, $h(t,x)$ is another two sided Brownian motion in x (plus a height shift). It is not the same Brownian motion. It is a new Brownian motion correlated with the first one in an extremely non-trivial way that is not understood (except the asymptotic correlations are known [**55**].)

At a completely formal level the invariance argument proceeds as follows. We work at the level of the stochastic Burgers equation on a large circle $[-L, L)$ with $L = -L$. Set $\lambda = \nu = \frac{1}{2}$, $D = 1$. As we pointed out, the Langevin equation $\partial_t u = \frac{1}{2}\partial_x^2 u + \partial_x \xi$ preserves white noise. So consider the Burgers flow $\partial_t u = \frac{1}{2}\partial_x(u^2)$. u lives in a space of rough functions $\mathscr{R}$ which we could take to be $H_{-1/2-\delta}$ for any $\delta > 0$. Let f be a nice function on $\mathscr{R}$. We hope to show that under the Burgers flow

$$\partial_t \int f(u(t))e^{-\frac{\sigma^2}{2}\int u^2} = 0 \tag{20}$$

where the integral is over $\mathscr{R}$ with respect to white noise with $E[u(t,x)u(s,y)] = \sigma^2\delta(t-s)\delta(x-y)$ which we write formally as $e^{-\frac{\sigma^2}{2}\int u^2}$. Differentiating we obtain

$$\partial_t \int f(u(t))e^{-\frac{\sigma^2}{2}\int u^2} = \int \langle \frac{\delta f}{\delta u}, \partial_t u\rangle e^{-\frac{\sigma^2}{2}\int u^2} \tag{21}$$

where $\frac{\delta f}{\delta u}$ is the functional (Frechet) derivative, and $\langle f,g\rangle = \int_{-L}^{L} fg dx$, to differentiate it from the function space integral. Now by the Burgers equation

$$\int \langle \frac{\delta f}{\delta u}, \partial_t u\rangle e^{-\frac{\sigma^2}{2}\int u^2} = \frac{1}{2}\int \langle \frac{\delta f}{\delta u}, \partial_x(u^2)\rangle e^{-\frac{\sigma^2}{2}\int u^2}. \tag{22}$$

Integrating by parts,

$$\frac{1}{2}\int \langle \frac{\delta f}{\delta u}, \partial_x(u^2)\rangle e^{-\frac{\sigma^2}{2}\int u^2} = -\frac{1}{2}\int f\langle \frac{\delta}{\delta u}\partial_x(u^2)e^{-\frac{\sigma^2}{2}\int u^2}\rangle. \tag{23}$$

But

$$\langle \frac{\delta}{\delta u}\partial_x(u^2)e^{-\frac{\sigma^2}{2}\int u^2}\rangle = \langle(2\partial_x u - \sigma^2 u\partial_x(u^2))\rangle e^{-\frac{\sigma^2}{2}\int u^2}. \tag{24}$$

The last term vanishes because $u\partial_x(u^2)_x = \partial_x\frac{1}{3}(u^3)$ and because of periodic boundary conditions any exact derivative integrates to zero: $\langle\partial_x f\rangle = \int_{-L}^{L}\partial_x f = 0$. This gives $\partial_t \int f(u(t))e^{-\frac{\sigma^2}{2}\int u^2} = 0$. So the Burgers part of the flow formally preserves white noise with *any* variance parameter σ^2. The constraint $\sigma = 1$ is set by the Langevin part.

Taking $L \to \infty$ gives the result on $\mathbb{R}$. Since the standard white noise with $\sigma^2 = 1$ is invariant for the Burgers flow $\partial_t u = \frac{1}{2}\partial_x(u^2)$ as well as the Langevin dynamics $\partial_t u = \frac{1}{2}\partial_x^2 u + \partial_x\xi$ it is invariant for the combined dynamics $\partial_t u = \frac{1}{2}\partial_x(u^2) + \frac{1}{2}\partial_x^2 u + \partial_x\xi$.

In this way, one concludes formally that white noise is invariant for the stochastic Burgers equation.

While the invariance is true, the above argument is not even correct at the physical level. The main problem is that the Burgers flow $\partial_t u = \frac{1}{2}\partial_x(u^2)$ is ill-defined (except for convex initial data) because there are no characteristics telling us how to fill in rarefaction fans. If one interprets the Burgers' flow as the usual entropy solutions, i.e. as the limit as $\nu \downarrow 0$ of $\partial_t u^\nu = u^\nu\partial_x u^\nu + \nu\partial_x^2 u^\nu$, then one has the Lax-Oleinik variational formula for the solution, $u = \partial_x h$ with

$$h(t,x) = \sup_{y\in\mathbb{R}}\left\{-\frac{(x-y)^2}{2t} + h(0,y)\right\}. \tag{25}$$

Starting from $h(0,x)$ a two-sided Brownian motion, one obtains a collection of 'N-waves', i.e. the indefinite space integral of a bunch of Dirac masses of various sizes, minus a linear function. The statistics are known exactly

[**62**] (see also [**17**], [**90**] for more general classes of solvable initial data). At any rate, the result starting with a Brownian motion is definitely *not* a new Brownian motion, though the formal argument tells you it should be.

A correct argument for the invariance is presented in Section 2.11 where it is proved by showing that KPZ is the weakly asymmetric limit of the height function of simple exclusion, which has symmetric random walks as its almost invariant measure. Alternatively, it could be obtained from the intermediate scaling limit of the special directed polymer model with log-Gamma distribution, which turns out to have product invariant measures for its free energy [**109**].

In general, one expects that for any initial data, no matter how smooth, the solution becomes locally Brownian at any positive time, with the same local diffusivity. (For proofs in various special cases, see [**103**], [**39**],[**67**], [**96**].)

1.5. Dynamic scaling exponent. We search for a scaling

$$h_\epsilon(t,x) = \epsilon^b h(\epsilon^{-z}t, \epsilon^{-1}x) \tag{26}$$

under which we can hope to see something non-trivial as $\epsilon \to 0$, i.e. on large space and time scales.

Either by taking a derivative, or subtracting constants independent of x, we can ignore global height shifts. Fixing $t = 0$, the fact that the solution is locally Brownian forces us to take

$$b = 1/2 \tag{27}$$

to see anything non-trivial. The equation becomes

$$\partial_t h_\epsilon = -\tfrac{1}{2}\epsilon^{2-z-b}(\partial_x h_\epsilon)^2 + \tfrac{1}{2}\epsilon^{2-z}\partial_x^2 h_\epsilon + \epsilon^{b-\frac{1}{2}z+\frac{1}{2}}\xi. \tag{28}$$

To avoid divergence of the nonlinear term we are then forced to take

$$z = 3/2. \tag{29}$$

In this way one arrives at the dynamic scaling exponent $3/2$.

All the universal fluctuation behaviour is observed under this scaling

$$h_\epsilon(t,x) = \epsilon^{1/2} h(\epsilon^{-3/2}t, \epsilon^{-1}x). \tag{30}$$

1.6. Renormalization of the nonlinear term. Once we know that Brownian motion is an invariant measure for the KPZ equation except for the height shift, i.e. in equilibrium $h(x) = h(t,x)$ is a two-sided Brownian motion in x, we realize the equation has a big problem. The term $(\partial_x h)^2$ cannot possibly make sense. Recall the basic quadratic variation computation for Brownian motion that for a real interval $[a,b]$, then, almost surely,

$$\lim_{n\to\infty} \sum_{i=\lfloor 2^n a\rfloor}^{\lfloor 2^n b\rfloor} |h(\tfrac{i+1}{2^n}) - h(\tfrac{i}{2^n})|^2 = (b-a). \tag{31}$$

The problem is not just because we started in equilibrium. The prediction is that starting with any, arbitrarily nice, initial data, for *any* time $t > 0$, the solution will be locally Brownian, i.e. (31) will hold.

We see that the non-linear term needs a kind of infinite renormalization. So it would be more honest to write the equation as[1]

$$\partial_t h = -[\tfrac{1}{2}(\partial_x h)^2 - \infty] + \tfrac{1}{2}\partial_x^2 h + \xi. \tag{32}$$

We have here a problem of scales in the derivation of the process. The nonlinear term is really being computed on a larger scale and is not supposed to be seeing the small scale fluctuations.

1.7. Cutoff KPZ models. What a stochastic partial differential equation is supposed to mean is that if we introduce some sort of cutoff, so that the equation makes classical sense, as we remove it we find an unambiguous limit, which is our solution. For example, in KPZ (1) we could smooth out the noise a little, with some sort of mollifier, to get ξ_ϵ. If $\xi_\epsilon(t,x)$ were now a smooth function of space and time, we could simply solve

$$\partial_t h_\epsilon = -\tfrac{1}{2}(\partial_x h_\epsilon)^2 + \tfrac{1}{2}\partial_x^2 h_\epsilon + \xi_\epsilon \tag{33}$$

for each realization of the noise separately. Then we try to take a limit as $\epsilon \to 0$ to get our solution. Alternatively, we could discretize the equation, for example in space, so that we have

$$\begin{aligned}\partial_t h(\tfrac{i}{n}) = -Q(h(\tfrac{i+1}{n}), h(\tfrac{i}{n}), h(\tfrac{i-1}{n}), h(\tfrac{i-2}{n})) + n^2(h(\tfrac{i+1}{n}) - 2h(\tfrac{i}{n})\\ + h(\tfrac{i-1}{n})) + n^{1/2} dB_{\frac{i}{n}}\end{aligned} \tag{34}$$

where $Q(h(\frac{i+1}{n}), h(\frac{i}{n}), h(\frac{i-1}{n}), h(\frac{i-2}{n}))$ is some sort of discretization of $\frac{1}{2}(\partial_x h)^2$. Another possibility is that h could be some sort of Markov chain or Markov process which might approximate h.

The last is in some sense the key point. The KPZ equation itself was built to model such systems, so it is crucial that whatever sense we try to make of it, it does model them correctly. Let us look at a few classic examples. All of them will be Markov processes on integer valued height functions h_i on a one dimensional lattice $\mathbb{Z}$.

- **Ballistic aggregation.** At each site one has "arrivals" occuring as a Poisson process with rate one. The arrivals at different sites are independent. When there is an arrival, the update rule is

$$h_i \mapsto \max\{h_{i-1}, h_i + 1, h_{i+1}\}. \tag{35}$$

  The arrivals can be thought of as particles which attach to the interface at the highest point at which there is either a particle to the right or left or below it.

[1]Before the mathematically inclined reader falls into despair, we should say that we will be completely precise about what we mean by all this later. At this point we are simply providing physical background to the problem.

- **Eden model.** We grow a finite connected subset A of $\mathbb{Z}^2$ by adding sites in the exterior boundary (i.e. sites in the complement of A which have a nearest neighbour in A). All such sites are added at rate one. To get a growing height function we let $h_i = \min\{j : (i, j+k) \notin A \text{ for all } k = 1, 2, \ldots\}$.
- **Restricted solid-on-solid/asymmetric exclusion.** The height profile is restricted to be a nearest neighbour random walk, i.e. $h_{i+1} - h_i \in \{-1, +1\}$. Each pair $(i, i+1)$ has two independent Poisson clocks which ring at rate p and at rate q. When the first clock rings, if the height differences there look like $(1, -1)$ we change it to $(-1, 1)$. Otherwise we do nothing. When the second clock rings, if the height differences there look like $(-1, 1)$ we change it to $(1, -1)$ and otherwise we do nothing. The model can also be encoded in terms of particle configurations on $\mathbb{Z}$, with at most one particle per site: A height difference of $+1$ corresponds to a particle, and a height difference of -1 corresponds to an empty site. The rule in this context is that particles perform a continuous time simple random walk, attempting to jump to the right at rate p and to the left at rate q. However, the jump takes place only if there is no particle at the target site, otherwise the particle keeps waiting where it is, and since we are in continuous time, we do not have to worry about ties. The case $p, q > 0$ with $p \neq q$ is called ASEP (Asymmetric Simple Exclusion Process) and $p = 0, q = 1$ is called TASEP (Totally Asymmetric Simple Exclusion Process).

The KPZ equation was introduced to model such growth processes, especially the ballistic aggregation (a nice introduction at this level is still [**69**].) In fact, to this day we know almost nothing about the fluctuations of the Eden model or ballistic aggregation. But the restricted solid on solid model/asymmetric exclusion process turns out to be solvable in a sense. Along with another solvable discretization called PNG (polynuclear growth model) [**77**], [**76**], [**101**] it has been the basis of the recent breakthroughs.

One reason (though not the only one) that asymmetric exclusion can be analyzed is that with appropriate boundary conditions its invariant measures are well understood. For example, if we put it on an infinite lattice, and have the flips happen as independent Poisson processes, the invariant measures are two sided random walks, which are discrete versions of Brownian motion.

For a discretization like (34), one has to be very careful with the choice of Q if one wants to have a nice invariant measure. For example, the very special discretization

$$Q(h_{i+1}, h_i, h_{i-1}, h_{i-2})) = (h_{i+1} - h_{i-2})(h_{i+1} - h_i - h_{i-1} + h_{i-2}) \tag{36}$$

works, in the sense that it preserves a random walk with Gaussian increments. This discretization was discovered by N.J. Zabusky [**94**] in the context of the Kortweg deVries equation (KdV) which bears many similarities to KPZ.

It is expected that if these cutoff versions of KPZ are not chosen carefully, they may not approximate KPZ. There could be no convergence, or convergence to a different answer. There are very few models in which we can prove any sort of convergence. They will be described in Section 2.11. For (36) we do not know the convergence, though it is expected to be correct (after subtraction of an appropriate diverging constant). For the invariant measures also, we only know them for a few special models. But the relation between having an explicit invariant measure and the convergence is also murky. It is not being used in what proofs we have in a way that is clear.

Another approach to understanding the KPZ equation is to introduce some sort of Wick ordered version of the nonlinearity,

$$:(\partial_x h)^2: \tag{37}$$

which is supposed to reflect what happens to $(\partial_x h)^2$ after the "∞" has been removed. Attempts in this direction (e.g. [**70**]) have unfortunately led to solutions which we know are non-physical because the scaling exponents are wrong [**31**][2].

Nevertheless, as we will see, there is a limiting object which in some sense satisfies the KPZ equation. The object is very canonical and we have even been able to obtain exact formulas for various quantities associated with it (see e.g. Theorem 1.3).

Perhaps a more straightforward approach would be to develop a theory for smoothed out noises. Everything would make sense. The problem is that there is some magic in the solution of KPZ and we would miss it this way. It is a classic mathematical situation in which one has to work hard to make sense of a canonical limiting problem, but then has an advantage of some exact solvability.

1.8. Hopf-Cole solutions. In the mid 90's, L. Bertini and G. Giacomin [**15**], following a suggestion of E. Presutti, proposed that the correct solutions of KPZ could be obtained as follows. The stochastic heat equation (with multiplicative noise) is

$$\partial_t z = \tfrac{1}{2}\partial_x^2 z - z\xi. \tag{38}$$

It has to be interpreted in the Itô sense, in which case it is well posed and for (reasonable) initial data $z_0(x) \geq 0$ with $\int z_0(x)dx > 0$ we have, for later times $t > 0$, $z(t,x) > 0$ for all x (see Theorem 2.11). Bertini and Giacomin proposed that

$$h(t,x) = -\log z(t,x) \tag{39}$$

[2]The reason is that their Wick products are based on the Gaussian structure of the forcing white noise ξ. The recent breakthrough work of Martin Hairer [**67**] uses the Gaussian structure of the Ornstein-Uhlenbeck process obtained by linearizing the KPZ equation, which does lead to the physical solutions.

is the correct solution of KPZ. There are multiple reasons to support this.

(1) If ξ were a nice function, then h would solve (1). This is just the classic Hopf-Cole transformation.

(2) Let's see what happens when ξ is a white noise. We start with the solution $z(t,x)$ of (38) and smooth it out a bit in space, then look for the equation it satisfies. For the computation we will need to assume we are in equilibrium. Let $\mathcal{G}_\kappa(x) = \frac{1}{\sqrt{2\pi\kappa^2}}\exp\{-x^2/2\kappa^2\}$ and

$$z_\kappa(t,x) = \langle \mathcal{G}_{x,\kappa}, z(t)\rangle = \int \mathcal{G}_\kappa(x-y)z(t,y)dy. \tag{40}$$

Define $h_\kappa(t,x) = -\log z_\kappa(t,x)$. Then by Itô's formula,

$$\partial_t h_\kappa + \tfrac{1}{2}(\partial_x h_\kappa)^2 - \tfrac{1}{2}\partial_x^2 h_\kappa - \xi = \left\{z_\kappa^{-1}\langle \mathcal{G}_{x,\kappa}z, \xi\rangle - \xi\right\} + \tfrac{1}{2}z_\kappa^{-2}\langle \mathcal{G}_{x,\kappa}^2, z^2\rangle. \tag{41}$$

For the first term, we can compute $E[(\int\int \varphi(t,x)\{z_\kappa^{-1}\langle \mathcal{G}_{x,\kappa}z,\xi\rangle - \xi\}dxdt)^2]$ for a smooth function φ of compact support by Itô isometry to get

$$\int\int E[(\int \tfrac{\varphi(t,y)\mathcal{G}_\kappa(y-x)dy}{\int \mathcal{G}_\kappa(y-y')z(t,y')dy'} z(t,x) - \varphi(t,x))^2]dxdt, \tag{42}$$

which vanishes as $\kappa \searrow 0$ by the continuity of $z(t,x)$.

We now compute the last term. Define $\mathcal{J}_\kappa(x) = 2\kappa\sqrt{\pi}\mathcal{G}_\kappa^2(x)$ so that $\mathcal{J}_\kappa$, $\kappa > 0$ is a new approximate identity. The last term is

$$\tfrac{1}{4}\kappa^{-1}\pi^{-1/2}\int \mathcal{J}_\kappa(x-y)e^{2(h(y)-h(x))}dy\ \left(\int \mathcal{G}_\kappa(x-y)e^{h(y)-h(x)}dy\right)^{-2}. \tag{43}$$

Because we are in equilibrium the $h(y)-h(x)$ are Brownian increments and we can make what is essentially a quadratic variation computation to get

$$\tfrac{1}{2}z_\kappa^{-2}\langle \mathcal{G}_{x,\kappa}^2, z^2\rangle \sim \tfrac{1}{4}\kappa^{-1}\pi^{-1/2}. \tag{44}$$

The precise constant $\frac{1}{2}\pi^{-1/2}$ just comes from the L^2 norm of the Gaussian smoothing. We could smooth with something else and get another constant.

So

$$\partial_t h_\kappa = -\tfrac{1}{2}[(\partial_x h_\kappa)^2 - \tfrac{1}{2}\kappa^{-1}\pi^{-1/2}] + \tfrac{1}{2}\partial_x^2 h_\kappa + \xi + o(1). \tag{45}$$

This gives our first precise form of (32).

(3) Suppose that instead of smoothing out z as above, we smooth out the white noise (in space),

$$\xi^\kappa(t,x) = \int \mathcal{G}_\kappa(x-y)\xi(t,y)dy.$$

Since the operation is linear $\xi^\kappa(t,x)$ is still Gaussian. It's mean is still zero and it has covariance

$$E[\xi^\kappa(t,x)\xi^\kappa(s,y)] = C_\kappa(x-y)\delta(t-s)$$

where

$$C_\kappa(x-y) = \int \mathcal{G}_\kappa(x-u)\mathcal{G}_\kappa(y-u)du.$$

In particular, we have again

(46) $$C_\kappa(0) = \tfrac{1}{2}\kappa^{-1}\pi^{-1/2}.$$

Let $z^\kappa(t,x)$ be the solution of the stochastic heat equation with the smoothed noise

(47) $$\partial_t z^\kappa = \partial_x^2 z^\kappa - z^\kappa \xi^\kappa, \qquad t>0, \quad x \in \mathbb{R}.$$

It is not difficult to show that $z^\kappa \to z$ uniformly on compact sets, and because $z(t,x)>0$ for $t>0$, we can define

(48) $$h^\kappa(t,x) = -\log z^\kappa(t,x).$$

and $h^\kappa(t,x)$ converge to $h(t,x) = -\log z(t,x)$. By Itô's formula, we have

(49) $$\partial_t h^\kappa = -\tfrac{1}{2}[(\partial_x h^\kappa)^2 - C_\kappa(0)] + \tfrac{1}{2}\partial_x^2 h^\kappa + \xi^\kappa.$$

Compare to (45).

(4) The Hopf-Cole solution is the one obtained by approximating KPZ by the free energy of directed random polymers in the intermediate disorder limit and by the height function of asymmetric exclusion [**15**] (see Section 2.11) in the weakly asymmetric limit. Such limits are expected to hold for a large class of systems with appropriate adjustable parameters. However, general proofs are not available at this time (see [**66**] for recent partial results).

(5) The Hopf-Cole solution has the conjectured scaling exponents. The first proof of this was [**11**] where it was shown that for the Hopf-Cole solution in equilibrium, i.e. starting from a two sided Brownian motion, there are $0 < C_1 \le C_2 < \infty$ such that

(50) $$C_1 t^{1/3} \le \sqrt{\mathrm{Var}(h(t,0))} \le C_2 t^{1/3}.$$

In these notes, we will also see that for special initial data one can go farther and obtain the conjectured asymptotic fluctuations: For the narrow wedge initial data (i.e. Dirac delta initial data for the stochastic heat equation) the F_{GUE} asymptotics were obtained in [**3, 108, 107**]; for half-Brownian initial data (i.e. initial data $e^{B(x)}1_{x>0}$ for the stochastic heat equation) the asymptotics were obtained in [**42**].

The evidence for the Hopf Cole solutions is now overwhelming. Whatever the physicists mean by KPZ, it is them. The problem of proving well-posedness for (1)-(3) is now seen to be of a very different nature from a problem like well-posedness for the 3-d incompressible Navier-Stokes equations. In the present case, we know that the solution is the Hopf-Cole solution. The problem is to find an appropriate definition of (1)-(3) which fits that solution, and to prove the corresponding uniqueness. As these notes were being produced, a solution to exactly this interpretation of the well-posedness problem has been announced by Martin Hairer [**67**][3].

Note that in higher dimensions one also has the formal Hopf-Cole formula linking

$$\partial_t h = -\tfrac{1}{2}|\nabla h|^2 + \tfrac{1}{2}\Delta h + \xi \tag{51}$$

with

$$\partial_t z = \tfrac{1}{2}\Delta z - z\xi \tag{52}$$

via

$$h(t,x) = -\log z(t,x). \tag{53}$$

Unfortunately, it can be checked that (52) does not have function valued solutions except in one dimension. So (53) does not make sense.

Discrete versions do hold. In particular, the partition functions of directed polymers in $d+1$ dimensions satisfy discrete versions of (52), and their logarithm (the free energy) satisfies a discrete version of (51). We now describe them.

1.9. Directed polymers in a random environment. The random environment is a collection $\xi(i,j)$ of independent identically distributed real random variables placed on the sites i,j of $\mathbb{Z}_+ \times \mathbb{Z}^d$. Given a realization $\xi(i,j)$ of the environment, the energy of an n-step nearest neighbour walk $\mathbf{x} = (x_1,\dots,x_n)$ is

$$H_n(\mathbf{x}) = \sum_{i=1}^{n} \xi(i,x_i).$$

Nearest neighbour walk just means a sequence of integers x_i, $i = 0,1,\dots,n$, with x_{i+1} an adjacent lattice site to x_i. The polymer measure on such walks starting at x at time 0 and ending at y at time n is then defined by

$$P_{x,n,y}(\mathbf{x}) = \frac{1}{Z(x,n,y)} e^{-\beta H_n(\mathbf{x})}\, P(\mathbf{x}),$$

where $\beta > 0$ is a parameter called the inverse temperature, P is the uniform probability measure on such walks, and $Z(x,n,y)$ is the partition function

$$Z(x,n,y) = \sum_{\mathbf{x}} e^{-\beta H_n^{\omega}(\mathbf{x})} P(\mathbf{x}) \tag{54}$$

This is called the point-to-point polymer. If we do not specify the end point, we call it the point-to-line polymer, and we call the partition function $Z(n)$. We will usually suppress the dependence on ξ and β except where

[3] Technically, the results of [**67**] are for KPZ on $[0,2\pi)$ with periodic boundary conditions, and extending them to $\mathbb{R}$ remains an open problem.

it may cause confusion, in which case we write things like $Z_\beta(n)$. For each realization of the environment we have a probability measure on random walk paths that prefers to travel through areas of low energy. Since the environment is random, we have a random probability measure on random walk paths.

They were introduced by [**71**] as a model of domain walls in Ising systems and arise as competition interfaces in multi-species growth [**68**]. There is a war between entropy (the number of such walks) and the energy. At $\beta = 0$ the polymer measure is of course just simple random walk, hence the walk is entropy dominated and exhibits diffusive behaviour. For β large, the polymer measure concentrates on special low energy paths which are no longer diffusive. Entropy domination is called *weak disorder*, and energy domination is called *strong disorder*. The precise separation between these two regimes is usually defined mathematically in terms of the positivity of the limit of the martingale $e^{-n\lambda(\beta)}Z(n)$ for the point-to-line partition function, as $n \to \infty$, where $\lambda(\beta) = E[e^{-\beta\omega}]$. The weak disorder regime consists of β for which

$$\lim_{n\to\infty} e^{-n\lambda(\beta)} Z_\beta(n) > 0,$$

whereas if the limit is zero then β is said to be in the strong disorder regime. For $d \geq 3$, it was shown early on [**73, 19**] that weak disorder holds for small β. Later, Comets and Yoshida [**36**] showed that in every dimension there is a critical value β_c such that weak disorder holds for $0 \leq \beta < \beta_c$ and strong disorder for $\beta > \beta_c$. In addition, for $d = 1$ and 2 they prove that $\beta_c = 0$. Understanding the polymer behaviour in the strong disorder regime is the main goal. The paths are superdiffusive with (point-to-line case)

$$|x(n)| \sim n^\zeta$$

for n large, with *transversal fluctuation exponent* $\zeta > 1/2$. For $d = 1$ the long-standing conjecture is $\zeta = 2/3$. For a long time, there were only non-sharp upper and lower bounds for special models [**35, 89, 99, 121, 122**]. Note that the picture is very different from simple random walk where the polymer endpoint is roughly uniformly distributed on an interval of size $\sqrt{n}$. For each realization of the random environment, the polymer is *localized* near a point of distance about n^ζ from the starting point. This point of course depends on the random environment. So the randomness in the polymer endpoint is basically a function of the random environment, and not the randomness of the random walk paths. Carmona and Hu [**29**] and Comets et al [**33**] showed that there is a constant $c_0 = c_0(\beta) > 0$ such that for the point-to-line

$$\overline{\lim_{n\to\infty}} \sup_{x\in\mathbb{Z}} P^\omega_{n,\beta}(x(n) = x) \geq c_0.$$

has probability one, in stark contrast to the simple random walk case where the supremum decays like $n^{-1/2}$. This is the best sort of rigorous mathematical statement of the localization that we have.

Strong disorder also can be seen in the large time behaviour of the partition function. There is a strict inequality

(55)
$$\rho(\beta) := \lim_{n\to\infty} \tfrac{1}{n}\log Z_\beta(n) = \lim_{n\to\infty} \tfrac{1}{n} E\log Z_\beta(n) < \lim_{n\to\infty} \tfrac{1}{n}\log EZ_\beta(n) := \lambda(\beta)$$

between the quenched and annealed free energies (the second equality is by a subadditivity argument and some concentration estimates, see for example [**29, 33**]). $\leq$ is obvious. The fact that it is strict was proved in $d = 1$ by Comets et al [**34**] (see also [**85**]). From (55) the leading term behavior of the log of the partition function is $\rho(\beta)n$.

The randomness is conjectured to appear through the random second order term

$$\log Z_\beta(n) \sim \rho(\beta)n + c(\beta)n^\chi X \tag{56}$$

where χ is the *longitudinal fluctuation exponent* and X represents asymptotic fluctuations of order 1. ζ and χ are supposed to satisfy the KPZ relation,

$$\chi = 2\zeta - 1. \tag{57}$$

Partial rigorous results were obtained recently by Chatterjee [**32**] (see also [**6**]).

In the following we will only be interested in $d = 1$. The conjecture is $\chi = 1/3$ and $\zeta = 2/3$. Results on $\zeta = 2/3$ have recently been obtained in special models [**11**],[**109**]. The connection to KPZ is made through the following discrete stochastic heat equation, satisfied by the partition function

$$Z(x, j+1, y) = \frac{1}{2}e^{-\beta\xi(j+1,y)}\left[Z(x, j, y+1) + Z(x, j, y-1)\right]. \tag{58}$$

1.10. Fluctuation breakthroughs of 1999. Now we turn to several models which are in some sense solvable and which are the source of all conjectures about the asymptotic fluctuations. Although these are truly mathematical results, we do not describe the mathematical details, but use the results to inspire the physical picture for KPZ.

Last passage percolation: Let $\xi(i,j)$, $(i,j) \in \mathbb{Z}_+^2$, be independent and identically distributed random variables. An *up/right path* π from $(1,1)$ to (M,N) is a sequence $(1,1) = (i_1, j_1), (i_2, j_2), \ldots, (i_m, j_m) = (M,N)$, $m = M + N - 1$, such that either $i_{r+1} - i_r = 1$ and $j_{r+1} = j_r$, or $i_{r+1} = i_r$ and $j_{r+1} - j_r = 1$. Set

$$G(M,N) = \max_\pi \sum_{(i,j)\in\pi} \xi(i,j), \tag{59}$$

where the maximum is taken over all up/right paths π from $(1,1)$ to (M,N).

One can see that last passage percolation is a kind of zero temperature $\beta \to \infty$ limit of the directed polymer in random environment, with $G(N, M)$ replacing the free energy, together with a 45° rotation of time and space. In fact it satisfies an equation,

$$G(M, N) = \max(G(M-1, N), G(M, N-1)) + \xi(M, N) \tag{60}$$

reminiscent of the Ballistic Aggregation model (35). This is another discrete version of the KPZ equation (not in the sense that it scales to the KPZ equation–there is no adjustable parameter–but in the sense that it is a discrete model with the same structure, and which is in the KPZ universality class.) If we had been careful, we could easily have defined discrete polymers in a way that the last passage percolation is *exactly* the zero temperature limit.

If $\xi(i, j)$ have the geometric distribution,

$$P(\xi(i, j) = k) = (1-q)q^k, \qquad k \in \{0, 1, 2, \ldots\}$$

this last passage percolation model can be seen as a discrete version of the polynuclear growth model (PNG) [**84**], [**100**]. Let $h(x, t) \in \mathbb{N}$ denote the height above $x \in \mathbb{Z}$ at time $t \in \mathbb{N}$. The model is defined by the discrete KPZ equation

$$h(x, t+1) = \max(h(x-1, t), h(x, t), h(x+1, t)) + \tilde{\xi}(x, t), \tag{61}$$

where $\xi(x, t)$ are independent random variables for $t \in \mathbb{Z}_+$ and $x \in \mathbb{Z}$ which vanish whenever $x - t$ is even, and have the geometric distribution when $x - t$ is odd. The connection between last passage percolation and this discrete PNG model is that if $\xi(i, j) = \tilde{\xi}(i - j, i + j - 1)$, then

$$G(i, j) = h(i - j, i + j - 1). \tag{62}$$

In this case we have a first example of an exact formula. K. Johansson [**76**] discovered that if $M \geq N$,

$$P(G(M, N) \leq s) = \det(I - K_N)_{L^2(s,\infty)}$$

where

$$K_N(x, y) = \frac{\kappa_{N-1}}{\kappa_N} \frac{p_N(x)p_{N-1}(y) - p_{N-1}(x)p_N(y)}{x - y}(w(x)w(y))^{1/2}. \tag{63}$$

$$w(x) = \binom{M - N + x}{x}$$

$p_N(x)$ are the Meixner polynomials, i.e. the normalized orthogonal polynomials $p_n(x) = \kappa_n x^n + \ldots$ with respect to the measure $w(x)dx$.

By taking appropriate asymptotic limits of this formula one obtains

$$G(N, N) \sim c_1 N + c_2 N^{1/3} \zeta \tag{64}$$

where ζ has the GUE Tracy-Widom distribution,

$$P(\zeta \leq s) = F_{\mathrm{GUE}}(s) = \det(I - K_{\mathrm{Ai}})_{L^2(s,\infty)} \tag{65}$$

where the *Airy kernel*

$$K_{\mathrm{Ai}}(x,y) = \int_{-\infty}^{0} \mathrm{Ai}(x-\lambda)\mathrm{Ai}(y-\lambda)d\lambda. \tag{66}$$

is the projection onto the negative eigenspace of the Airy operator

$$H = -\partial_x^2 + x \tag{67}$$

and $\det(I+A)_{L^2(\Omega,d\mu)}$ is the Fredholm determinant of the operator A with integral kernel $A(x,y)$ on $L^2(\Omega,d\mu)$ (i.e. $Af(x) = \int_\Omega A(x,y)f(y)d\mu(y)$) defined by the formula of H. von Koch,

$$\det(I+A) = 1 + \sum_{n=1}^{\infty} \frac{1}{n!} \int_{\Omega^n} \det(A(x_i,x_j))_{i,j=1}^{n} \prod_{i=1}^{n} d\mu(x_i) \tag{68}$$

which was used by I. Fredholm to establish the criteria $\det(I+A) \neq 0$ for solvability of $(I+A)f = g$, and is now called the *Fredholm expansion* (see [**13**]). A must be compact, though it need not be symmetric for this to make sense. The series on the right hand side does converge whenever A is trace-class, i.e.

$$\|A\|_1 = \mathrm{tr}(|A|) < \infty \tag{69}$$

where $|A| = \sqrt{A^*A}$, in which case, as expected,

$$\det(I+A) = \prod_n (1+\lambda_n(A)) \tag{70}$$

where $\lambda_n(A)$ are the eigenvalues of A, counted with multiplicity [**110**].

The remarkable thing is that the fluctuations are the same as those of the largest eigenvalue $\lambda_N^{\max}$ of a matrix from the Gaussian unitary ensemble (GUE), i.e. a Hermetian $N \times N$ matrix $a_{ij} = \overline{a_{ji}}$, $i,j = 1,\dots,N$ such that for $i < j$, a_{ij} is distributed as $\mathscr{N}(0,\sqrt{N}/2) + i\mathscr{N}(0,\sqrt{N}/2)$ and on the diagonal $i = j$, a_{ii} is distributed as $\mathscr{N}(0,\sqrt{N})$ where $\mathscr{N}(a,b)$ means Gaussian (=normal) mean 0 and variance b. The a_{ii} and the real and imaginary parts of the a_{ij} are all independent. The $\sqrt{N}$ is an arbitrary normalization which makes the analogy with growth models more transparent. Alternatively, the probability measure on the space of Hermitian matrices is

$$Z_N^{-1} e^{-\frac{1}{2N}\mathrm{Tr}A^2} \prod_{i=1}^{N} da_{ii} \prod_{i<j=1}^{N} d\mathrm{Re}a_{ij} d\mathrm{Im}a_{ij}. \tag{71}$$

By the Wigner semicircle law, the spectrum has approximately a semicircle density on $[-2N, 2N]$ and

$$\lambda_N^{\max} \sim 2N + N^{1/3}\zeta \tag{72}$$

where ζ has the same GUE Tracy-Widom distribution.

That the fluctuations of a random growth model are the same as those of the top eigenvalue of a GUE matrix came as a surprise. New results came quickly on a number of solvable models, and physicists predicted sweeping generalizations of the fluctuation results to the KPZ universality class.

In the totally asymmetric case $p = 0, q = 1$ [**76**] (and much later in the asymmetric case $p > 0, p + q = 1$ [**116**]) the fluctuations of asymmetric exclusion were also shown to be asymptotically $F_{\rm GUE}$: *Let $q - p \in (0, 1]$ and start with $h(0, x) = |x|$, then*

$$h(\tfrac{t}{q-p}, 0) \sim \tfrac{t}{2} - 2^{-1/3}t^{1/3}\zeta \tag{73}$$

where ζ is GUE Tracy-Widom. Note that as usual what we mean by this expression is really $\lim_{t\to\infty} P\big(\frac{h(\frac{t}{q-p},0)-\frac{t}{2}}{2^{-1/3}t^{1/3}} \geq -s\big) = F_{\rm GUE}(s)$.

In fact, the result mentioned above for geometric last passage percolation followed an earlier result [**7**] for the longest increasing subsequence of a random permutation, which can be thought of as a growth model in an analogous way. In this context J. Baik and E. Rains studied the analogue of the point-to-line polymer, and discovered that the fluctuations were also of size $N^{1/3}$, but now governed asymptotically by the $F_{\rm GOE}$ Tracy-Widom distribution for the Gaussian Orthogonal Ensemble,

$$F_{\rm GOE}(s) = \det(I - P_0 B_s P_0)_{L^2(\mathbb{R})} \tag{74}$$

where

$$B_m(x, y) = \mathrm{Ai}(x + y + m) \tag{75}$$

and P_a is the projection onto $L^2(a, \infty)$ [**54**].

If we consider the largest eigenvalue $\lambda_N^{\max}$ of a matrix from the Gaussian orthogonal ensemble (GOE), i.e. the probability measure on the space of real symmetric matrices is

$$Z_N^{-1} e^{-\frac{1}{2N}\mathrm{Tr}A^2} \prod_{i\leq j=1}^{N} da_{ij}. \tag{76}$$

As before we have

$$\lambda_N^{\max} \sim 2N + N^{1/3}\zeta \tag{77}$$

but now ζ has the $F_{\rm GOE}$ distribution.

Adjusting boundary conditions so that they correspond to being flat on one side and curved on the other, Baik and Rains obtain a family of distributions interpolating between $F_{\rm GUE}$ distribution and the $F_{\rm GOE}$ distribution. Starting with equilibrium (Brownian like) boundary conditions, they discover a new universal distribution which they call F_0 (see [**10**]).

Here we have only scratched the surface of the solvable models. For nice review articles the reader is directed to [**53**], [**56**], [**78**], especially for the relation between last passage percolation and the RSK (Robinson-Schensted-Knuth) correspondence. A very recent article [**41**] discusses many such exact formulas in the context of the tropical version of RSK (see also [**21**] for very general results about solvability in this context.)

All the one-point distributions were then generalized to many space points. The resulting universal spatial fluctuation processes are called the Airy processes.

1.11. The Airy processes. These stochastic processes are defined through their finite dimensional distributions, which are given by Fredholm determinants. Note that this is quite different from being a determinental process. However, there is some connection. The Airy processes are derived from exact formulas for TASEP and PNG which come from them being parts of determinental processes. In particular, they are top lines of non-intersecting line ensembles, whose determinental structure comes from the Karlin-McGregor formula.

The Airy$_2$ process $\mathcal{A}_2(x)$, $x \in \mathbb{R}$. Given $\lambda_0, \dots, \lambda_n \in \mathbb{R}$ and $x_0 < \cdots < x_n$ in $\mathbb{R}$, the finite dimensional distributions are given by

(78)
$$P(\mathcal{A}_2(x_0) \le \lambda_0, \dots, \mathcal{A}_2(x_n) \le \lambda_n) = \det(I - \mathrm{f}^{1/2} K_{\mathrm{ext}} \mathrm{f}^{1/2})_{L^2(\{x_0,\dots,x_n\}\times\mathbb{R})},$$

where we have counting measure on $\{x_0, \dots, x_n\}$ and Lebesgue measure on $\mathbb{R}$, f is defined on $\{x_0, \dots, x_n\} \times \mathbb{R}$ by $\mathrm{f}(x_j, \lambda) = \mathbf{1}_{\lambda \in (\lambda_j, \infty)}$, and the *extended Airy kernel* [**102, 59, 87**] is defined by

$$K_{\mathrm{ext}}(x, \xi; x', \xi') = \begin{cases} \int_0^\infty ds\, e^{-s(x-x')} \mathrm{Ai}(\xi + s)\mathrm{Ai}(\xi' + s), & \text{if } x \ge x' \\ \int_{-\infty}^0 ds\, e^{-s(x-x')} \mathrm{Ai}(\xi + s)\mathrm{Ai}(\xi' + s), & \text{if } x < x', \end{cases}$$

The Airy$_2$ process is the dynamical version of F_{GUE} in the following sense. The Dyson Brownian motion is a stationary time dependent version of the ensemble which has the GUE as its distribution at every time. The matrices evolve according to the Ornstein-Uhlenbeck process given by the Langevin equation,

$$dA(x) = -\tfrac{1}{2N} A(x)dx + dB(x) \tag{79}$$

where $B_{ii}(x)$, $i = 1, \dots, N$, $\mathrm{Re}B_{ij}(x)$, $\mathrm{Im}B_{ij}$, $i < j = 1, \dots, N$ are independent Brownian motions, the first with diffusion coefficient 1 and the latter two with diffusion coefficients 1/2. The largest eigenvalue $\lambda_N^{\max}(x)$ of $A(x)$ is now a function of x and

$$\lambda_N^{\max}(t) \sim 2N + N^{1/3} \mathcal{A}_2(x). \tag{80}$$

A generalization has the sequence of x dependent eigenvalues converging to a non-intersecting line ensemble called the *Airy line ensemble* [**40**]. It was originally derived from the PNG model by M. Prähofer and H. Spohn [**101**] (see also [**77**]). There is also a variant of (78) due to [**102**],

(81)
$$P(\mathcal{A}_2(x_0) \le \lambda_0, \dots, \mathcal{A}_2(x_n) \le \lambda_n) = \det\Big(I - K_{\mathrm{Ai}} + \bar{P}_{\lambda_0} e^{(x_0 - x_1)H} \bar{P}_{\lambda_1} e^{(x_1 - x_2)H} \cdots \bar{P}_{\lambda_n} e^{(x_n - x_0)H} K_{\mathrm{Ai}}\Big),$$

$\bar{P}_a = 1 - P_a$ is the projection onto $L^2(-\infty, a]$. Airy$_2$ governs the asymptotic spatial fluctuations in growth models starting from curved initial data and point-to-point random polymers.

The Airy$_1$ process $\mathcal{A}_1(x)$ is another stationary process, whose one-point distribution is now F_{GOE}. It is defined through its finite-dimensional distributions, given by a determinantal formula: for $\lambda_1, \ldots, \lambda_n \in \mathbb{R}$ and $x_1 < \cdots < x_n$ in $\mathbb{R}$,

(82)
$$P(\mathcal{A}_1(x_1) \le \lambda_1, \ldots, \mathcal{A}_1(x_n) \le \lambda_n) = \det(I - \mathrm{f}^{1/2} K_1^{\mathrm{ext}} \mathrm{f}^{1/2})_{L^2(\{x_1,\ldots,x_n\}\times\mathbb{R})},$$

where we have counting measure on $\{x_1, \ldots, x_n\}$ and Lebesgue measure on $\mathbb{R}$, f is defined on $\{x_1, \ldots, x_n\} \times \mathbb{R}$ by $\mathrm{f}(x_j, \lambda) = \mathbf{1}_{\lambda \in (\lambda_j, \infty)}$ and

$$K_1^{\mathrm{ext}}(x, \xi; x', \xi') = -\frac{1}{\sqrt{4\pi(x'-x)}} \exp\left(-\frac{(\xi'-\xi)^2}{4(x'-x)}\right) \mathbf{1}_{x'>x} \tag{83}$$

$$+\mathrm{Ai}(\xi + \xi' + (x'-x)^2) \exp\left((x'-x)(\xi+\xi') + \frac{2}{3}(x'-x)^3\right). \tag{84}$$

The finite-dimensional distributions of the Airy$_1$ process are also given by the following formula: for $\lambda_1, \ldots, \lambda_n \in \mathbb{R}$ and $x_1 < \cdots < x_n$ in $\mathbb{R}$,

(85)
$$P(\mathcal{A}_1(x_1) \le \lambda_1, \ldots, \mathcal{A}_1(x_n) \le \lambda_n)$$
(86)
$$= \det\left(I - B_0 + \bar{P}_{\lambda_1} e^{-(x_1-x_2)\Delta} \bar{P}_{\lambda_2} e^{-(x_2-x_1)\Delta} \cdots \bar{P}_{\lambda_n} e^{-(x_n-x_1)\Delta} B_0\right)_{L^2(\mathbb{R})}.$$

Note however, that it is not true that the Airy$_1$ process is the limit of the largest eigenvalue process in the matrix valued diffusion for GOE [**20**] (and it is an open problem what is.) Airy$_1$ arises in growth models starting from flat initial data and point-to-line random polymers [**105**],[**23**].

The Airy$_{\mathrm{stat}}$ process $\mathcal{A}_{\mathrm{stat}}(x)$ is the equilibrium analogue of the Airy$_1$ and Airy$_2$ processes. It has finite dimensional distributions [**8**]

$$\begin{aligned} &P(\mathcal{A}_{\mathrm{stat}}(x_1) \le \lambda_1, \ldots, \mathcal{A}_{\mathrm{stat}}(x_n) \le \lambda_n) \\ &= \sum_{k=1}^{n} \frac{\partial}{\partial \lambda_k} \left(g_n(x, \lambda) \det\left(I - \mathrm{f}^{1/2} \widehat{K}_{\mathrm{Ai}} \mathrm{f}^{1/2}\right)_{L^2(\{x_1,\ldots,x_n\}\times\mathbb{R})}\right) \end{aligned} \tag{87}$$

where

$$\widehat{K}_{\mathrm{Ai}}((i, \xi), (j, \xi'))$$
(88)
$$= \begin{cases} \int_0^{\infty} ds \mathrm{Ai}(\xi + s + x_i^2) \mathrm{Ai}(\xi' + s + x_j^2) e^{-s(x_j - x_i)}, & \text{if } x_i \le x_j, \\ -\int_{-\infty}^{0} ds \mathrm{Ai}(\xi + s + x_i^2) \mathrm{Ai}(\xi' + s + x_j^2) e^{-s(x_j - x_i)}, & \text{if } x_i > x_j. \end{cases}$$

The function $g_n(x,\lambda)$ is defined by

$$g_n(x,\lambda) = \mathcal{R} + \sum_{i=1}^{n}\sum_{j=1}^{n}\int_{\lambda_i}^{\infty} du \int_{\lambda_j}^{\infty} du'\, \Psi_j(u')\rho_{j,i}(u',u)\Phi_i(u), \tag{89}$$

where

$$\rho := (I - \mathrm{f}^{1/2}\widehat{K}_{\mathrm{Ai}}\mathrm{f}^{1/2})^{-1}, \qquad \rho_{j,i}(u',u) := \rho((j,u'),(i,u)), \tag{90}$$

and $\Phi((i,\lambda)) := \Phi_i(\lambda)$, $\Psi((j,\lambda)) = \Psi_j(\lambda)$. The functions $\mathcal{R}$, Φ, and Ψ are defined by

$$\mathcal{R} = \lambda_1 + e^{-\frac{2}{3}x_1^3}\int_{\lambda_1}^{\infty} du \int_0^{\infty} du'\, \mathrm{Ai}(u+u'+x_1^2)e^{-x_1(u+u')},$$

$$\Psi_j(\lambda) = e^{\frac{2}{3}x_j^3 + x_j\lambda} - \int_0^{\infty} du\, \mathrm{Ai}(u+\lambda+x_j^2)e^{-x_j u},$$

$$\begin{aligned}\Phi_i(\lambda) = e^{-\frac{2}{3}x_1^3}\int_0^{\infty} du \int_{\lambda_1}^{\infty} du'\, e^{-u(x_1-x_i)}e^{-x_1 u'}\mathrm{Ai}(\lambda+x_i^2+u)\\ \times \mathrm{Ai}(u'+x_1^2+u) + \mathbf{1}_{[i\geq 2]}\frac{e^{-\frac{2}{3}x_i^3 - x_i\lambda}}{\sqrt{4\pi(x_i-x_1)}}\int_{-\infty}^{\lambda_1-\lambda} du'\, e^{-\frac{(u')^2}{4(x_i-x_1)}}\\ - \int_0^{\infty} du'\, \mathrm{Ai}(u'+\lambda+x_i^2)e^{x_i u'}.\end{aligned} \tag{91}$$

for $i,j = 1,2,\dots,n$. The marginal $\mathcal{A}_{\mathrm{stat}}(0)$ has the Baik-Rains distribution F_0 [**9**], [**55**]. Unlike the previous two, the Airy$_{\mathrm{stat}}$ process is not stationary. In fact, because of the invariance of Brownian motion, it is nothing but a two sided Brownian motion with a height shift $\mathcal{A}_{\mathrm{stat}}(0)$. The complication, of course, is that the height shift and the Brownian motion are highly dependent.

These are the three most basic Airy processes, but there are are three more obtained from data which is, say curved to one side of the origin and flat on the other side (the Airy$_{2\to 1}$ process $\mathcal{A}_{2\to 1}(x)$), or curved on one side and Brownian on the other (the Airy$_{2\to BM}$ process $\mathcal{A}_{2\to BM}(x)$), or flat on one side and Brownian on the other (the Airy$_{1\to BM}$ process $\mathcal{A}_{1\to BM}(x)$). They look like one of the three basic ones far to one side, and the other far to the other side, e.g. $\mathcal{A}_{2\to 1}(x) \to \mathcal{A}_2(x)$ as $x \to -\infty$ and $\mathcal{A}_{2\to 1}(x) \to \mathcal{A}_1(x)$ as $x \to \infty$, and the one dimensional distributions $F_x(s) = P(\mathcal{A}_{2\to 1}(x) \leq s)$ interpolates between F_{GUE} and F_{GOE}. The other two look like Brownian motion with a non-trivial height shift asymptotically in one direction, but this is not really $\mathcal{A}_{\mathrm{stat}}(x)$ because the height shift there is very particular. Of course, because of the statistical $x \mapsto -x$ symmetry of KPZ, or the stochastic heat equation, if one starts, say, with initial data curved on the right hand side and flat one the left, one expects asymptotically to have spatial fluctuations $\mathcal{A}_{2\to 1}(-x)$. The finite dimensional distributions of the mixed Airy processes are also given by Fredholm determinants (those for Airy$_{2\to BM}$ can be found in [**72**], for Airy$_{2\to 1}$ in [**24**], and for Airy$_{1\to BM}$ in

[25], in which it is the process $\mathcal{A}_{2\to 1,M,\kappa S_\nu}$ with $M = 1$ and $\kappa = 0$, the single slow particle simulating Bernoulli 1/2 to the right of the origin and flat to the left.)

1.12. Predicted fluctuations for the KPZ universality class. Extrapolating from the exact results for TASEP and PNG, the conjectural picture that has developed is that the universality class is divided into sub-universality classes which depend on the initial data class, but not on other details of the particular models. Because of their self-similarity properties, the three basic initial data are, at the level of continuum partition functions (taking logarithms gives free energies or height functions): Dirac δ_0, corresponding to curved, or droplet type initial data; 0, corresponding to growth off a flat substrate; and $e^{B(x)}$ where $B(x)$ is a two sided Brownian motion, corresponding to growth in equilibrium. Of course, in discrete models of various types one is dealing with discrete approximations of such initial data. There are also three additional non-homogeneous sub-universality classes corresponding to starting with one of the basic three on one side of the origin, and another on the other side. The spatial fluctuations in these six basic classes of initial data are supposed to be given asymptotically by the six known Airy processes.

We will state the conjectures in terms of the Hopf-Cole solution $h(t,x) = -\log z(t,x)$ of KPZ where $z(t,x)$ is the solution of the stochastic heat equation (38) with prescribed initial data. This is because the initial data are most transparent in that context. All the models in the universality class have an analogue of the height function for which the conjectures are expected to hold.

<u>1. *Curved*</u>, corresponding to initial data $z(0,x) = \delta_0$, i.e. KPZ starting from *narrow wedge initial condition*,

$$-2^{1/3}t^{-1/3}(h(t,2^{1/3}t^{2/3}x) - 2^{-1/3}t^{1/3}x^2 - \tfrac{t}{24} - \log\sqrt{2\pi t}) \to \mathcal{A}_2(x). \tag{92}$$

More generally, it will hold for KPZ growth with curved initial data and point-to-point directed random polymers, with new non-universal constants. The factors $2^{1/3}$ represent a the change from the standard choice of normalization in random matrices to the standard choice in growth models. The non-universal constants can be computed in cases where the flux function is known explicitly [111]. So far, this is only the case in models with explicit invariant measures.

<u>2. *Flat*</u>, corresponding to initial data $z(0,x) = 1$, i.e. KPZ starting from $h(0,x) = 0$,

$$-2^{1/3}t^{-1/3}(h(t,2^{1/3}t^{2/3}x) - \tfrac{t}{24}) \to \mathcal{A}_1(x). \tag{93}$$

More generally, it will hold for KPZ growth on asymptotically flat substrates, point-to-line directed random polymers, etc.

3. *Equilibrium*, corresponding to $z(0,x) = e^{B(x)}$ where $B(x)$ is a two-sided Brownian motion,

$$-2^{1/3}t^{-1/3}(h(t, 2^{1/3}t^{2/3}x) - \tfrac{t}{24}) \to \mathcal{A}_{\mathrm{stat}}(x). \tag{94}$$

There are completely analogous conjectures in the three transitional cases. Since we will use it, we state the $\mathrm{Airy}_{2\to BM}$ case.

4. *Half-Brownian*, corresponding to initial data $z(0,x) = 1_{x>0}e^{B(x)}$, i.e. KPZ starting from *half-Brownian initial conditions*,

$$-2^{1/3}t^{-1/3}(h(t, 2^{1/3}t^{2/3}x) - 2^{-1/3}t^{1/3}x^2 - \tfrac{t}{24}) \to \mathcal{A}_{2\to BM}(x). \tag{95}$$

All the conjectures are known only in a few special solvable models with determinental structure, even at the level of one-point distributions. Therefore it was a considerable surprise when exact results involving Fredholm determinants became available for the asymmetric exclusion process, which is *not* of this type.

1.13. Tracy-Widom formulas for ASEP. In a series of articles starting with [**114, 115**] C. Tracy and H. Widom developed exact contour integral formulas for the transition probabilities of ASEP with a finite number of particles (ASEP is described in detail in Section 2.14.) The formula for n particles is written in terms of an n-fold contour integral. It was a great surprise that in the special case where one starts with all sites to the left of the origin initially empty, and with Bernoulli product measure, density ρ to the right of the origin, one obtains an exact formula for the one-point distribution of the height function, involving a Fredholm determinant. The first formula was for the case $\rho = 1$, i.e. with initial height function $h(0,x) = |x|$, called the *corner growth model*. The second formula was for the case $\rho < 1$, called *step Bernoulli*. We state them together. Recall that $h(t,x)$, $t \geq 0$, $x \in \mathbb{Z}$ is the height function for ASEP with local minima jumping to local maxima at rate q and local maxima jumping to local minima at rate p.

THEOREM 1.1. [**116**],[**117**] *Let $q > p$ with $q + p = 1$, $\gamma = q - p$, $\tau = p/q$, $\alpha = (1-\rho)/\rho$. For $m = \lfloor \frac{1}{2}(s+x) \rfloor$, $t \geq 0$ and $x \in \mathbb{Z}$*

$$P(h(t,x) \geq s) = \int_{S_{\tau+}} \frac{d\mu}{\mu} \prod_{k=0}^{\infty} (1 - \mu\tau^k) \det(I + \mu J_\mu)_{L^2(\Gamma_\eta)} \tag{96}$$

where $S_{\tau+}$ is a circle centered at zero of radius strictly between τ and 1, and where the kernel of the determinant is given by

$$J_\mu(\eta, \eta') = \int_{\Gamma_\zeta} \exp\{\Lambda(\zeta) - \Lambda(\eta')\} \frac{f(\mu, \zeta/\eta')}{\eta'(\zeta - \eta)} \frac{g(\eta')}{g(\zeta)} d\zeta \tag{97}$$

where

$$
\begin{aligned}
f(\mu, z) &= \sum_{k=-\infty}^{\infty} \frac{\tau^k}{1-\tau^k\mu} z^k \\
\Lambda(\zeta) &= -x\log(1-\zeta) + \frac{t\zeta}{1-\zeta} + m\log\zeta \\
g(\zeta) &= \prod_{n=0}^{\infty}(1+\tau^n\alpha\zeta).
\end{aligned} \tag{98}
$$

The contours are as follows: η and η' are on Γ_η, a circle symmetric about the real axis and intersecting it at $-\alpha^{-1}+2\delta$ and $1-\delta$ for δ small. And the ζ integral is on Γ_ζ, a circle of diameter $[-\alpha^{-1}+\varsigma, 1+\varsigma]$. One should choose ς so as to ensure that $|\zeta/\eta| \in (1, \tau^{-1})$. This choice of contour avoids the poles of g which are at $-\alpha^{-1}\tau^{-n}$ for $n \geq 0$. In the corner growth case this last point is unnecessary as $g(\eta')/g(\zeta) = 1$.

Note that the formulas are very similar in structure, except that the step Bernoulli case has the extra $\frac{g(\eta')}{g(\zeta)}$.

It was known [**15**] that the Hopf-Cole solutions of KPZ could be obtained by diffusively rescaling ASEP with q very close to p. Hence, results for KPZ could be lifted from ASEP results if one could do the appropriate asymptotics. All exact results for KPZ that are known have been obtained in this fashion.

1.14. Exact results for KPZ. We now state some of the mathematical results we can obtain for the KPZ equation. The first one did not use exact formulas, and only gives the size of fluctuations in equilibrium.

THEOREM 1.2 (Equilibrium. [**11**]). *Let $z(t,x)$ be the solution of the stochastic heat equation (38) with initial data $z(0,x) = e^{B(x)}$. Let $h(t,x) = -\log z(t,x)$ be the corresponding solution of KPZ. There are constants $0 < c_1 < c_2 < \infty$ such that*

$$
c_1 t^{2/3} \leq \mathrm{Var}(h(t,0)) \leq c_2 t^{2/3}. \tag{99}
$$

The proof of Theorem 1.2 proceeds via formulas which relate the variance of the height function to the variance of a second class particle in the exclusion processes. These are then studied using coupling arguments. The main problem is to keep all the estimates appropriately uniform in the weakly asymmetric limit when q and p are close. We will not go into details here. The interested reader can consult [**11**] and references therein.

The following exact formula is obtained by studying the weakly asymmetric limit of the Tracy-Widom formula for the corner growth model by steepest descent.

THEOREM 1.3. **[3]**, **[108]** *Let $z(t,x)$ be as in the previous theorem and let $h(t,x) = -\log z(t,x)$ and define*

$$F_t(s) = P(h(t,x) + \tfrac{x^2}{2t} + \log\sqrt{2\pi t} + \tfrac{t}{24} \geq -s). \tag{100}$$

$F_t(s)$ does not depend on x and is given by the crossover formula

$$F_t(s) = \int_C \frac{d\mu}{\mu} e^{-\mu} \det(I - K_{\sigma_{t,\mu}})_{L^2(\kappa_t^{-1}s,\infty)} \tag{101}$$

where $\kappa_t = 2^{-1/3}t^{1/3}$, C is a contour positively oriented and going from $+\infty + \epsilon i$ around $\mathbb{R}^+$ to $+\infty - i\epsilon$, and K_σ is an operator given by its integral kernel

(102)

$$K_\sigma(x,y) = \int_{-\infty}^{\infty} \sigma(\tau)\mathrm{Ai}(x+\tau)\mathrm{Ai}(y+\tau)d\tau, \qquad \textit{and} \qquad \sigma_{t,\mu}(\tau) = \frac{\mu}{\mu - e^{-\kappa_t\tau}}.$$

The last formula was derived independently and at the same time by the two groups using the same method. **[3]** provided a complete proof. **[108]** derived the exact formula, and provided numerical plots of the new distribution functions. In addition, they compute the long-time correction to F_{GUE} and provide extensive physical context. However, at several key points in the argument, they proceed without mathematical justification. First in the weakly asymmetric limit, where existing results of **[15]** did not apply in the needed context. In fact, a logarithmic shift is necessary, as will be explained later in these notes. **[108]** determine the needed size of the shift by matching first moments, but do not provide further justification. In particular, they do not show that the resulting distributions are non-degenerate. The second key point is that **[108]** deform contours during the asymptotic analysis in order to end up with a nice formula (the limiting contours should be the nice ones for the Airy functions). However, during the asymptotics, a diverging number of residues builds up as poles pass through these contours. The resulting correction to the formula should, in principle, be enormous. They are ignored in **[108]**. Seredipidously, the sum of poles turns out to vanish. But we only knows this because these issues were addressed rigorously, at the same time, in the complete proof of **[3]**.

Next we turn to the case of half Brownian initial data, as in (95). We can prove this in the sense of one dimensional marginals, by taking the $t \to \infty$ limit in another surprising exact formula.

THEOREM 1.4. **[42]** *Let $h(t,x)$ be the Hopf-Cole solution of KPZ starting from half-Brownian initial data. Let*

$$F_{t,x}^{\mathrm{edge}}(s) := P\left(\frac{h(t, 2^{1/3}t^{2/3}x) - \frac{t}{24}}{t^{1/3}} \geq 2^{-1/3}x^2 - s\right). \tag{103}$$

Then

$$F_{t,x}^{\mathrm{edge}}(s) = \int_{\tilde{\mathcal{C}}} e^{-\tilde{\mu}} \frac{d\tilde{\mu}}{\tilde{\mu}} \det(I - K_s^{\mathrm{edge}})_{L^2(\tilde{\Gamma}_\eta)}.$$

The contour $\tilde{\mathcal{C}}$ is given as

$$\tilde{\mathcal{C}} = \{e^{i\theta}\}_{\pi/2 \le \theta \le 3\pi/2} \cup \{x \pm i\}_{x>0}$$

The contours $\tilde{\Gamma}_\eta$, $\tilde{\Gamma}_\zeta$ are given as

$$\tilde{\Gamma}_\eta = \{2^{-7/3} + ir : r \in (-\infty, -1) \cup (1, \infty)\} \cup \tilde{\Gamma}_\eta^d \tag{104}$$

$$\tilde{\Gamma}_\zeta = \{-2^{-7/3} + ir : r \in (-\infty, -1) \cup (1, \infty)\} \cup \tilde{\Gamma}_{\eta}^d, \tag{105}$$

where $\tilde{\Gamma}_\zeta^d$ is a dimple which goes to the right of $xt^{-1/3}$ and joins with the rest of the contour, and where $\tilde{\Gamma}_\eta^d$ is the same contour just shifted to the right by distance $2^{-4/3}$. The kernel K_s^{edge} acts on the function space $L^2(\tilde{\Gamma}_\eta)$ through its kernel:

$$\begin{aligned} &K_s^{\mathrm{edge}}(\tilde{\eta}, \tilde{\eta}') \\ &= \int_{\tilde{\Gamma}_\zeta} \exp\{-\frac{t}{3}(\tilde{\zeta}^3 - \tilde{\eta}'^3) + st^{1/3}(\tilde{\zeta} - \tilde{\eta}')\} \\ &\quad \times \frac{\pi 2^{1/3}(-\tilde{\mu})^{-2^{1/3}(\tilde{\zeta}-\tilde{\eta}')}}{\sin(\pi 2^{1/3}(\tilde{\zeta} - \tilde{\eta}'))} \frac{\Gamma(2^{1/3}\tilde{\zeta} - 2^{1/3}xt^{-1/3})}{\Gamma(2^{1/3}\tilde{\eta}' - 2^{1/3}xt^{-1/3})} \frac{d\tilde{\zeta}}{\tilde{\zeta} - \tilde{\eta}} \end{aligned} \tag{106}$$

where $\Gamma(z)$ is the Gamma function.

In Section 2.16 we describe the steepest descent analysis necessary to obtain the previous two results from the Tracy-Widom formulas. From the exact formulas an easy asymptotics gives

COROLLARY 1.5. [**3, 42**] *(92) and (95) hold in the sense of converges of one-dimensional distributions.*

In the first case, this resolves the basic conjecture for KPZ that for the narrow wedge initial data, for each x, $-2^{1/3}t^{-1/3}(h(t,x) - x^2/2t - \frac{t}{24} - \log\sqrt{2\pi t})$ converges in distribution to F_{GUE}, which is a refined statement that the KPZ equation is indeed in the KPZ universality class.

In a remarkable concurrent development, the F_{GUE} and F_{GOE} fluctuations were observed in experiments of nematic liquid crystals by K.A. Takeuchi and M. Sano [**112, 113**].

1.15. KPZ universality, or universality of KPZ?. All of the models described in this section, as well as the KPZ equation, are believed to be members of the *KPZ universality class*, in the sense that they should have the scaling exponent $z = 3/2$, and, at a more refined level, the correct fluctuations (GUE/Airy$_2$, GOE/Airy$_1$, Baik-Rains/Brownian motion) at the scale (30) depending on the initial conditions (curved, flat, equilibrium). We have sketched above what is proved for special cases.

There is another, much more restrictive, type of universality, that of the KPZ equation itself. As we will see, it arises when one takes a continuum limit of cutoff models, while making at the same time a critical adjustment of the non-linearity. Not all models have such an adjustable non-linearity. Asymmetric exclusion does, the parameter being $q-p$. There is also a large class of generalized asymmetric exclusion models, which do [**66**]. All directed polymers do, the parameter being β. LPP does not, and, as far as we can tell, PNG does not. These limits are described in Section 2.11. The exact results for KPZ then mean that we can obtain weak versions of some of the universality conjectures (see Corollary 2.20). This is the advantage of the KPZ equation over other models in the universality class.

1.16. KPZ fixed point. All the fluctuations, for all these models, are observed under the rescaling,

$$h_\epsilon(t,x) = (R_\epsilon h)(t,x) := \epsilon^{1/2} h(\epsilon^{-3/2}t, \epsilon^{-1}x) \tag{107}$$

after subtraction of appropriately diverging quantities.

If we started with the standard KPZ, this would satisfy the KPZ with renormalized coefficients,

$$\partial_t h_\epsilon = \tfrac{1}{2}(\partial_x h_\epsilon)^2 + \epsilon^{1/2}\tfrac{1}{2}\partial_x^2 h_\epsilon + \epsilon^{1/4}\xi. \tag{108}$$

Now we ask what happens as $\epsilon \to 0$. The limiting process should contain all the fluctuation behaviour we have observed so far. It is the fixed point of the renormalization (107). Presumably, it contains a lot of the integrable structure. So far, we do not know much about it. The rest of this section consists of conjectures [**43**].

We define the *KPZ fixed point* $\mathfrak{h}$ as the $\epsilon \to 0$ limit of the properly centered process $\bar{h}_\epsilon$. Note that we do not know in general that such a limit exists[4]. We now give a conjectural construction of the limiting object.

Let $h(u,y;t,x)$ be the solution of (1) for times $t > u$ started at time u with a delta function at y, all using the same noise. To center, set $\bar{h}(u,y;t,x) = h(u,y;t,x) - \frac{t-u}{24} - \log\sqrt{2\pi(t-u)}$ and define A_1 by

$$\bar{h}(u,y;t,x) = -\tfrac{(x-y)^2}{2(t-u)} + A_1(u,y;t,x).$$

After the rescaling (107),

$$R_\epsilon \bar{h}(u,y;t,x) = -\tfrac{(x-y)^2}{2(t-u)} + A_\epsilon(u,y;t,x)$$

where $A_\epsilon = R_\epsilon A_1$. The R_ϵ acts on the two pairs of variables here. As $\epsilon \to 0$, $A_\epsilon(u,y;t,x)$ converges to the *Airy sheet* $\mathcal{A}(u,y;t,x)$. In each spatial variable it is an Airy_2 process. It has several nice properties:

(1) *Independent increments.* $\mathcal{A}(u,y;t,x)$ is independent of $\mathcal{A}(u',y;t',x)$ if $(u,t)\cap(u',t') = \emptyset$;

(2) *Space and time stationarity.* $\mathcal{A}(u,y;t,x) \overset{\text{dist}}{=} \mathcal{A}(u+h,y;t+h,x) \overset{\text{dist}}{=} \mathcal{A}(u,y+z;t,x+z)$;

[4] Partial results have been obtained for Poissonian last passage percolation [**30**]. There one can obtain the necessary tightness, which means that such an object exists. What is missing is its uniqueness, and, more generally the universality.

(3) *Scaling.* $\mathcal{A}(0,y;t,x) \overset{\text{dist}}{=} t^{1/3}\mathcal{A}(0,t^{-2/3}y;1,t^{-2/3}x)$;

(4) *Semi-group property.* For $u < s < t$,

$$\mathcal{A}(u,y;t,x) = \sup_{z\in\mathbb{R}}\Big\{\tfrac{(x-y)^2}{2(t-u)} - \tfrac{(z-y)^2}{2(s-u)} - \tfrac{(x-z)^2}{2(t-s)} + \mathcal{A}(u,y;s,z) + \mathcal{A}(s,z;t,x)\Big\}. \tag{109}$$

Using $\mathcal{A}(u,y;t,x)$ we construct the KPZ fixed point $\mathfrak{h}(t,x)$. By the Hopf-Cole transformation and the linearity of the stochastic heat equation, the rescaled solution of (1) with initial data h^0 is

$$R_\epsilon h(t,x) = \epsilon^{1/2}\ln\int e^{\epsilon^{-1/2}\{-\frac{(x-y)^2}{2t} + A_\epsilon(0,y;t,x) + R_\epsilon h^0(y)\}}dy. \tag{110}$$

If we choose initial data h^0_ϵ so that $R_\epsilon h^0_\epsilon$ converges to a fixed function f in the limit, we can use Laplace's method to evaluate $\mathfrak{h}(t,x) = \lim_{\epsilon\to 0} R_\epsilon h(t,x) = T_{0,t}f(x)$ where

$$T_{u,t}f(x) := \sup_y\Big\{-\tfrac{(x-y)^2}{2(t-u)} + \mathcal{A}(u,y;t,x) + f(y)\Big\}. \tag{111}$$

The operators $T_{u,t}$, $0 < u < t$ form a semi-group, i.e. $T_{u,t} = T_{u,s}T_{s,t}$ which is stationary with independent increments and $T_{0,t} \overset{\text{dist}}{=} R^{-1}_{t^{-2/3}}T_{0,1}R_{t^{-2/3}}$.

By the Markov property, the joint distribution of the marginal spatial process of $\mathfrak{h}$ (for initial data f) at a set of times $t_1 < t_2 < \cdots < t_n$ is given by

$$(\mathfrak{h}(t_1),\ldots,\mathfrak{h}(t_n)) \overset{\text{dist}}{=} (T_{0,t_1}f,\ldots,T_{t_{n-1},t_n}\cdots T_{0,t_1}f).$$

The process of randomly evolving functions can be thought of as a high dimensional analogue of Brownian motion (with state space Brownian motions!), and the $T_{t_i,t_{i+1}}$ as analogous to the independent increments.

The solution of (1) corresponds to the free energy of a directed random polymer $x(s)$, $u < s < t$ starting at y and ending at x, with quenched random energy

$$H(x(\cdot)) = \textstyle\int_u^t\{|\dot{x}(s)|^2 - \xi(s,x(s))\}ds. \tag{112}$$

Under the rescaling (7) this probability measure on paths converges to the *polymer fixed point*; a continuous path $\pi_{u,y;t,x}(s)$, $u \le s \le t$ from y to x which at discrete times $u = s_0 < \cdots s_{m-1} < t$ is given by the argmax over $x_0,\ldots,x_{m-1}$ of

$$(T_{u,s_1}\delta_y)(x_1) + (T_{s_1,s_2}\delta_{x_1})(x_2) + \cdots + (T_{s_{m-1},t}\delta_{x_{m-1}})(x). \tag{113}$$

This is the analogue in the present context of the minimization of the action and the polymer fixed point paths are analogous to characteristics in the randomly forced Burger's equation. One might hope to take the analogy farther and find a limit of the renormalizations of (112), and minimize it to find that path $\pi_{u,y;t,x}$. However, the limit does not appear to exist, so one has to be satisfied with the limiting paths themselves. The path $\pi_{0,y;t,x}$ turns out to be Hölder continuous with exponent $1/3-$, as compared to Brownian motion where the Hölder exponent is $1/2-$. As the mesh of times is made finer, a limit $\mathcal{E}(\pi_{0,y;t,x})$ of (113) does exist, and through it we can write the time evolution of the KPZ fixed point in terms of the polymer fixed point through the analogue of the Lax-Oleinik variational formula,

$$\mathfrak{h}(t,x) = \sup_{y\in R}\{\mathcal{E}(\pi_{0,y;t,x}) + f(\pi_{0,y;x,t}(0))\}. \tag{114}$$

The KPZ fixed point, Airy sheet, and polymer fixed point are universal and will arise in random polymers, last passage percolation and growth models – anything in the KPZ universality class. Just as for (1), at the microscopic scale, approximate versions of the variational problem (111) hold, becoming exact as $\epsilon \to 0$.

2. A mathematical introduction

In the following sections we give a rigorous introduction to white noise and the stochastic heat equation. This will give us a chance to introduce many of the key concepts in stochastic pde (for alternate surveys see **[97]**, **[45]**.) Following that we sketch the proofs that discrete models converge to the stochastic heat equation, and the corresponding limit of the Tracy-Widom step Bernoulli formula, which then leads to the edge crossover distributions. So we now switch from heuristics to rigorous mathematics.

2.1. White noise and stochastic integration in 1+1 dimensions. White noise $\xi(t,x)$, $t\geq 0$, $x\in\mathbb{R}$ is the distribution valued Gaussian process with mean zero and covariance

$$E[\xi(t,x)\xi(s,y)] = \delta(t-s)\delta(x-y). \tag{115}$$

More precisely, for any smooth function f of compact support we can write[5]

$$\int_{\mathbb{R}_+\times\mathbb{R}} f(t,x)\xi(t,x)dxdt, \tag{116}$$

and the random variables $\{\int_{\mathbb{R}_+\times\mathbb{R}} f(t,x)\xi(t,x)dxdt\}_{f\in L^2(\mathbb{R}_+\times\mathbb{R})}$ are jointly Gaussian with mean zero and covariance

$$\begin{aligned} &E[\int_{\mathbb{R}_+\times\mathbb{R}} f_1(t,x)\xi(t,x)dxdt \int_{\mathbb{R}_+\times\mathbb{R}} f_2(t,x)\xi(t,x)dxdt] \\ &= \int_{\mathbb{R}_+\times\mathbb{R}} f_1(t,x)f_2(t,x)dxdt. \end{aligned} \tag{117}$$

There are many ways to construct it. One way is to choose an orthonormal basis $f_1, f_2, \ldots$ of $L^2(\mathbb{R}_+\times\mathbb{R})$, and independent Gaussian random variables $Z_1, Z_2, \ldots$, each with mean zero and variance one, and write

$$\xi(t,x) = \sum_{n=1}^{\infty} Z_n f_n(t,x). \tag{118}$$

[5]We will always use the natural notation $\int fg$ when f is a smooth test function and g is a distribution.

One can check the resulting object makes sense as an element of the negative Sobolev space $H_{-1-\delta,\mathrm{loc}}(\mathbb{R}_+ \times \mathbb{R})$ for $\delta > 0$.

We assume that our white noise has been constructed on a probability space $(\Omega, \mathscr{F}, P)$. This could be in the way described above, or some other way. We now start to construct stochastic integrals. For non-random functions this is easy. We have done it in (116) for smooth functions of compact support in $\mathbb{R}_+ \times \mathbb{R}$. If $f \in L^2(\mathbb{R}_+ \times \mathbb{R})$ then we define

$$\int_{\mathbb{R}_+ \times \mathbb{R}} f(t,x)\xi(t,x)dxdt \tag{119}$$

by approximation. Indeed, there are smooth functions f_n of compact support in $\mathbb{R}_+ \times \mathbb{R}$, with

$$\int_{\mathbb{R}_+ \times \mathbb{R}} |f_n(t,x) - f(t,x)|^2 dxdt \to 0. \tag{120}$$

The f_n are therefore Cauchy in $L^2(\mathbb{R}_+ \times \mathbb{R})$. But by (117) we have

$$E[(\int_{\mathbb{R}_+ \times \mathbb{R}} (f_n(t,x) - f_m(t,x))\xi(t,x)dxdt)^2] = \int_{\mathbb{R}_+ \times \mathbb{R}} |f_n(t,x) - f_m(t,x)|^2 dxdt \tag{121}$$

so $\int_{\mathbb{R}_+ \times \mathbb{R}} f_n(t,x)\xi(t,x)dxdt$ is a Cauchy sequence in $L^2(\Omega, \mathscr{F}, P)$, and therefore has a limit. Hence (119) exists as a limiting object in $L^2(\Omega, \mathscr{F}, P)$ because of the completeness of that space.

However, we will also need to integrate random functions, in particular functions which depend on the white noise ξ. It is here that things become a little more complicated, and, as in the one dimensional case, one has to make a choice of integral.

We now define the stochastic integral we need. Because we are dealing with a parabolic type stochastic partial differential equation, the time and space directions are treated differently. Our stochastic integral is basically the standard Itô integral, but only in the time variable. As in the one dimensional case we start with the simplest integrators and build the integral by approximation.

For smooth functions φ on $\mathbb{R}$ with compact support and $t > 0$ we can define

$$\int_{\mathbb{R}_+ \times \mathbb{R}} \mathbf{1}_{(0,t]}(s)\varphi(x)\xi(s,x)dxds. \tag{122}$$

For each fixed φ, it is a Brownian motion in t, with variance $\int \varphi^2(x)dx$, i.e. the Gaussian process with mean zero and covariance from (117),

$$E[\int \mathbf{1}_{(0,t_1]}(s)\varphi(x)\xi(s,x)dxds \int \mathbf{1}_{(0,t_2]}(s)\varphi(x)\xi(s,x)dxds]$$
$$= \min(t_1,t_2)\int \varphi^2(x)dx. \tag{123}$$

Let $\mathscr{F}_0 = \emptyset$ and for each $t > 0$, define $\mathscr{F}_t$ to be the σ-field generated by

$$\left\{\int \mathbf{1}_{(0,s]}(u)\varphi(x)\xi(u,x)dxdu\right\},$$
$$0 \le s \le t,\ \varphi \text{ a smooth function of compact support on } \mathbb{R}.$$

$\mathscr{F}_s \subseteq \mathscr{F}_t$ if $s \le t$, i.e. $\mathscr{F}_t$ is a filtration in $\mathscr{F}$.

Let $\mathcal{S}$ be the set of functions of the form

$$f(t,x,\omega) = \sum_{i=1}^{n} X_i(\omega)\mathbf{1}_{(a_i,b_i]}(t)\varphi_i(x) \tag{124}$$

where X_i is a bounded random variable measurable with respect to $\mathscr{F}_{a_i}$, $0 \le a_i \le b_i < \infty$ and φ_i are smooth function of compact support on $\mathbb{R}$. For $f \in \mathcal{S}$ we define the stochastic integral as

$$\int_{\mathbb{R}_+\times\mathbb{R}} f(t,x)\xi(t,x)dxdt = \sum_{i=1}^{n} X_i \int_{\mathbb{R}_+\times\mathbb{R}} \mathbf{1}_{(a_i,b_i]}(t)\varphi_i(x)\xi(t,x)dxdt. \tag{125}$$

the important point is that the time increment sticks out into the future. For such integrators one easily checks the linearity of the integral and the isometry

$$E\left[\left(\int_{\mathbb{R}_+\times\mathbb{R}} f(t,x)\xi(t,x)dxdt\right)^2\right] = \int_{\mathbb{R}_+\times\mathbb{R}} E[f^2(t,x)]dxdt. \tag{126}$$

Let $\mathscr{P}$ denote the sub-σ-field of $\mathscr{B}(\mathbb{R}_+\times\mathbb{R})\times\mathscr{F}$ generated by $\mathcal{S}$. Let $L^2(\mathbb{R}_+\times\mathbb{R}\times\Omega,\mathscr{P})$ be the linear subspace of $f(t,x,\omega)$ measurable with respect to $\mathscr{P}$ and square integrable. These will be our integrators. Like all Itô integrals the key point is that they are in some sense "non-anticipating", i.e. $f(t,x,\omega)$ only depends on the information $\mathscr{F}_t$ up to time t. We construct the stochastic integral for them through the isometry, by approximation, so we need the following

LEMMA 2.1. *$\mathcal{S}$ is dense in $L^2(\mathbb{R}_+\times\mathbb{R}\times\Omega,\mathscr{P})$.*

PROOF. The proof is the same as the one dimensional case. If $f \in L^2(\mathbb{R}_+\times\mathbb{R}\times\Omega,\mathscr{P})$, then $f_n = f\mathbf{1}_{|f|\le n, t\le n}$ are bounded, supported on $t \le n$ and measurable with respect to $\mathscr{P}$ and $\|f - f_n\|_{L^2(\mathbb{R}_+\times\mathbb{R}\times\Omega,\mathscr{P})} \to 0$ by the monotone convergence theorem. If f is bounded and measurable with respect to $\mathscr{P}$, and supported on $t \le t_0 < \infty$ then

$$f_n(t,x,\omega) = \sum_{k=0}^{\infty} \mathbf{1}_{[\frac{k}{2^n},\frac{k+1}{2^n})}(s)\frac{1}{2^n}\int_{\frac{k-1}{2^n}}^{\frac{k}{2^n}} f(s,x,\omega)ds \tag{127}$$

are in $\mathcal{S}$ and $\|f - f_n\|_{L^2(\mathbb{R}_+\times\mathbb{R}\times\Omega,\mathscr{P})} \to 0$ by the Lebesgue differentiation and bounded convergence theorems. □

Consequently, if $f \in L^2(\mathbb{R}_+ \times \mathbb{R} \times \Omega, \mathscr{P})$ we can choose $f_n \in \mathcal{S}$ with $f_n \to f$ in[($L^2(\mathbb{R}_+ \times \mathbb{R} \times \Omega, \mathscr{P})$. By the isometry (126),

$$I_n(\omega) = \int_{\mathbb{R}_+ \times \mathbb{R}} f_n(t, x, \omega)\xi(t, x, \omega)dxdt$$

is a Cauchy sequence in $L^2(P)$, i.e. $\lim_{n,m\to\infty} E[(I_n - I_m)^2] = 0$. Hence there is a limit point $I \in L^2(P)$ which we call the stochastic integral $\int_{\mathbb{R}_+ \times \mathbb{R}} f(t, x, \omega)\xi(t, x, \omega)dxdt$. It is linear in f and we have the Itô isometry (126).

For future use we state

LEMMA 2.2 (Burkholder's inequality). *For $p \geq 2$ there is a $C_p < \infty$ such that*

(128)

$$E[|\int_{\mathbb{R}_+ \times \mathbb{R}} f(t, x, \omega)\xi(t, x, \omega)dxdt|^p] \leq C_p E[|\int_{\mathbb{R}_+ \times \mathbb{R}} f^2(t, x, \omega)dxdt|^{p/2}]$$

Burkholder's inequality also holds for discrete approximations of the stochastic integral. In this context (which we will use) it goes back to Khinchine, and Marcinkiewicz-Zygmund, whose version is for independent, mean zero random variables X_i, $i = 1, 2, \ldots$ with finite pth moment, stating that $E[|\sum_{i=1}^n X_i|^p] \leq C_p E[(\sum_{i=1}^n X_i^2)^{p/2}]$.

Sketch of proof of Lemma 2.2. Apply Itô's lemma with the function $|x|^p$ to the martingale $M_t := \int_0^t \int f\xi dxds$ to obtain

$$E[|M_t|^p] = \tfrac{p(p-1)}{2} E\left[\int_0^t |M_s|^{p-2} \int f^2(s, x)dxds\right]. \tag{129}$$

Call $\bar{M}_{t,p} = E[\sup_{0\leq s\leq t} |M_t|^p]$. By Doob's maximal inequality,

$$\bar{M}_{t,p} \leq (\tfrac{p}{p-1})^p E[|M_t|^p]. \tag{130}$$

Applying Hölder's inequality to the right side of (129) we obtain, from (130),

$$\bar{M}_{t,p} \leq c_p(\bar{M}_{t,p})^{\frac{p-2}{p}} E\left[\left(\int_0^t \int f^2 dxds\right)^{\frac{p}{2}}\right]^{\frac{2}{p}}, \tag{131}$$

which gives Lemma 2.2. Note that for a complete proof, one has to approximate $|x|^p$ by a function with two bounded derivatives to make the application of Itô's equation valid. We leave the details as an exercise for the reader. □

2.2. Wiener chaos. We can also define multiple stochastic integrals,

(132)
$$\int_{\Lambda_k}\int_{\mathbb{R}^k} f(t_1,\ldots,t_k,x_1\ldots,x_k)\xi(t_1,x_1)\cdots\xi(t_k,x_k)dx_1\cdots dx_k dt_1\cdots dt_k$$
$$=\int_{\Lambda_k}\int_{\mathbb{R}^k} f(\mathbf{t},\mathbf{x})\xi^{\otimes\mathbf{k}}(\mathbf{t},\mathbf{x})\mathbf{dt}\,\mathbf{dx}$$

where $\Lambda_k=\{0\le t_1<t_2<\cdots<t_n<\infty\}$. For example, in the previous section we have constructed $\int_0^{t_2}\int_{\mathbb{R}} f(t_1,t_2,x_1,x_2)\xi(t_1,x_1)dx_1dt_1$ and shown that it is progressively measurable. Hence we can define $\int_{0\le t_1<t_2<\infty}\int_{\mathbb{R}^2} f(t_1,t_2,x_1,x_2)\xi(t_1,x_1)\xi(t_2,x_2)dx_1dt_1dt_2$ as $\int_0^\infty\int_{\mathbb{R}}\Big[\int_0^{t_2}\int_{\mathbb{R}} f(t_1,t_2,x_1,x_2)\,\xi(t_1,x_1)dx_1dt_1\Big]dx_2dt_2$. The construction extends inductively to (132). The resulting multiple stochastic integrals have

$$E\left[\int_{\Lambda_k}\int_{\mathbb{R}^k} f(\mathbf{t},\mathbf{x})\xi^{\otimes\mathbf{k}}(\mathbf{t},\mathbf{x})\mathbf{dt}\,\mathbf{dx}\int_{\Lambda_\mathbf{j}}\int_{\mathbb{R}^\mathbf{j}} \mathbf{g}(\mathbf{t},\mathbf{x})\xi^{\otimes\mathbf{j}}(\mathbf{t},\mathbf{x})\mathbf{dt}\,\mathbf{dx}\right]$$
$$=\langle f,g\rangle_{L^2(\Lambda_k\times\mathbb{R}^k)}\mathbf{1}_{j=k}. \tag{133}$$

and in fact they span $L^2(\Omega,\mathscr{F},P)$ (see, e.g. **[75]** for a nice treatment).

In this way, there is an isometry between $L^2(\Omega,\mathscr{F},P)$ and $\bigoplus_{k=0}^\infty L^2(\Lambda_k\times\mathbb{R}^k)$, i.e. given a random variable $X\in L^2(\Omega,\mathscr{F},P)$, there are $f_k\in L^2(\Lambda_k\times\mathbb{R}^k), k=0,1,2,\ldots$, with

$$X=\sum_{k=0}^\infty\int_{\Lambda_k}\int_{\mathbb{R}^k} f_k(\mathbf{t},\mathbf{x})\xi^{\otimes\mathbf{k}}(\mathbf{t},\mathbf{x})\mathbf{dt}\,\mathbf{dx}.$$

Here f_0 is simply the constant EX, with the convention $\int f_0=f_0$. By (133),

$$E[X^2]=\sum_{k=0}^\infty||f_k||^2_{L^2(\Lambda_k\times\mathbb{R}^k)}.$$

This Wiener chaos representation can be very useful. But the problem is that if φ is non-linear, then knowing the chaos representation of X tells us nothing about the chaos representation of $\varphi(X)$. For example, in Section 2.3 we will see that the chaos series for the solution $z(t,x)$ of the stochastic heat equation is particularly simple. But unfortunately, we do not know the chaos series for the main quantity of interest in these notes, $\log z(t,x)$, the Hopf-Cole solution of KPZ.

2.3. The stochastic heat equation. The stochastic heat equation is

$$\partial_t z=\tfrac{1}{2}\partial_x^2 z - z\xi,\qquad t>0,\ x\in\mathbb{R},$$
$$z(0,x)=z_0(x), \tag{134}$$

where $\xi(t,x)$ is space-time white noise. z_0 could be taken to be random, but we will always assume that it and the white noise ξ are independent.

We will usually assume that $z_0(x) \geq 0$. In that case, as long as z_0 is not identically zero, we will prove in Theorem 2.11 that $z(t,x) > 0$, in which case $h(t,x) = -\log z(t,x)$ is well defined. Recall that $h(t,x)$ is the Hopf-Cole solution of KPZ (1).

All our physical problems will only involve z_0 which are positive in some sense. However, for purposes of proving uniqueness we will have to take differences. The solvability will depend on the initial data, and will have to include objects like exponential Brownian motions, as well as delta function initial data. A reasonable class which includes everything we want is $z_0(x)$ such that $\int_A z_0(x)dx$ is a signed measure in A satisfying for some $c < \infty$,

$$E[\sup_{A\subset[-n,n]} \int_A z_0(x)dx] < ce^{cn}. \tag{135}$$

This class is preserved by the heat semigroup.

2.4. Mild solutions. To make sense of the equation, we rewrite it in Duhamel form

$$z(t,x) = \int_{\mathbb{R}} p(t,x-y)z_0(y)dy - \int_0^t \int_{\mathbb{R}} p(t-s,x-y)z(s,y)\xi(s,y)dyds \tag{136}$$

using the kernel of the heat equation

$$p(t,x) = \frac{1}{\sqrt{2\pi t}}e^{-x^2/2t}. \tag{137}$$

If $z(t,x)$ is progressively measurable with

$$\int_0^t \int_{\mathbb{R}} p^2(t-s,x-y)E[z^2(s,y)]dyds < \infty \tag{138}$$

then the stochastic integral in (136) makes sense as an element of $L^2(P)$ as shown in the previous section. Such a $z(t,x)$ for which equality holds in (136) for all $0 \leq t \leq T$ and $x \in \mathbb{R}$ will be called a *mild solution* of the stochastic heat equation (134).

Of course we need to know such a solution exists. There are many ways to do it, but an easy way is Picard iteration: Let $z^0(t,x) = 0$ and

$$z^{n+1}(t,x) = \int_{\mathbb{R}} p(t,x-y)z_0(y)dy - \int_0^t \int_{\mathbb{R}} p(t-s,x-y)z^n(s,y)\xi(s,y)dyds. \tag{139}$$

They are progressively measurable by construction. Let $\bar{z}^n(t,x) = z^{n+1}(t,x) - z^n(t,x)$. Then

$$\bar{z}^{n+1}(t,x) = -\int_0^t \int_{\mathbb{R}} p(t-s,x-y)\bar{z}^n(s,y)\xi(s,y)dyds. \tag{140}$$

So

$$E[|\bar{z}^{n+1}(t,x)|^2] = \int_0^t \int_{\mathbb{R}} p^2(t-s,x-y)E[|\bar{z}^n(s,y)|^2]dyds. \tag{141}$$

Let $f^n(t) = \sup_{x,s\in[0,t]} E[|\bar{z}^n(s,x)|^2]$. Suppose that $f^0(t) < \infty$. Then we have, after integrating the heat kernel,

$$f^{n+1}(t) \le C \int_0^t \frac{f^n(s)}{\sqrt{t-s}} ds. \tag{142}$$

Iterating the inequality,

$$f^{n+1}(t) \le C \int_0^t \int_0^s \frac{f^{n-1}(u)}{\sqrt{(t-s)(s-u)}} duds. \tag{143}$$

Changing the order of integration

$$f^{n+1}(t) \le C \int_0^t \int_u^t \frac{f^{n-1}(u)}{\sqrt{(t-s)(s-u)}} dsdu = C' \int_0^t f^{n-1}(u)du. \tag{144}$$

Now Gronwall's inequality implies that $f^n(t) \le (C't)^{n/2}/(n/2)!$, so there is a limit $z(t,x)$ which is progressively measurable and solves the equation.

If we examine what $f^0(t) < \infty$ means, it is just that $\sup_x E[z_0^2(x)] < \infty$. This is pretty restrictive. In fact, typical initial data we are interested in are (1) $z_0(x) = \delta_0(x)$ (2) $z_0(x) = e^{B(x)} 1_{x>0}$. Both lie out of this class. Thus we need an a priori estimate for such solutions.

LEMMA 2.3. *Let $z(t,x)$ satisfy (136) with z_0 as in (135). Then there exists a $c < \infty$ for all $x \in \mathbb{R}$,*

$$E[z^2(t,x)] \le ct^{-1/2} e^{c(t+|x|)}. \tag{145}$$

In particular, let $z(t,x)$ satisfy (136) with $z_0 = \delta_0$ (or, in general, localized initial data). Then there exists a $c = c(T)$ such that for all $0 < t \le T$, for all

$$E[z^2(t,x)] \le cp^2(t,x). \tag{146}$$

PROOF. Without loss of generality we can assume the initial data is non-random because we could always take the conditional expectation given $\mathscr{F}_0$, and then take a further expectation to get the result. Then, squaring (136) and taking expectation we have

(147)

$$E\left[z^2(t,x)\right] = \left(\int_{\mathbb{R}} p(t,x-y)z_0(y)dy\right)^2 + \int_0^t \int_{\mathbb{R}} p^2(t-s,x-y)E\left[z^2(s,y)\right] dyds.$$

Iterating the equation, we obtain

$$E\left[z^2(t,x)\right] = \sum_{n=0}^{\infty} E[I_n(t,x)] \tag{148}$$

where, for $\Delta_n(t) = \{0 \le t_n \le t_{n-1} \le \cdots \le t_0 = t\}$, $x_0 = x$

$$I_n(t,x) = \int_{\Delta_n(t)} \int_{\mathbb{R}^n} \prod_{i=0}^{n-1} p^2(t_i - t_{i+1}, x_i - x_{i+1})$$

$$\times \left(\int_{\mathbb{R}} p(t_n, x_n - y) z_0(y) dy \right)^2 \prod_{i=1}^{n} dx_i dt_i. \tag{149}$$

Now we use the fact that for $s < u < t$

$$\int_{\mathbb{R}} p^2(t-u, x-z) p^2(u-s, z-y) dz = \sqrt{\tfrac{t-s}{4\pi(t-u)(u-s)}} p^2(t-s, x-y) \tag{150}$$

to simplify

$$I_n(t,x) = \frac{1}{\Gamma(\frac{n-1}{2})} \int_{\Delta_n(t)} \prod_{i=1}^{n} \frac{dt_i}{\sqrt{4\pi(t_{i-1} - t_i)}} \int_{\mathbb{R}} p(\tfrac{t_0 - t_n}{2}, x_0 - x_n)$$

$$\times \left(\int_{\mathbb{R}} p(t_n, x_n - y) z_0(y) dy \right)^2 dx_n$$

$$= \frac{2^{-n+\frac{1}{2}}}{\Gamma(\frac{n-1}{2})} \int_0^t \int (t-s)^{\frac{n}{2}} p^2(t-s, x-z)$$

$$\times \left(\int_{\mathbb{R}} p(s, z-y) z_0(y) dy \right)^2 dz ds$$

$$\le C(\tfrac{t}{4})^{n/2} (n!)^{-1/2} \int_0^t \int p^2(t-s, x-z)$$

$$\times \left(\int_{\mathbb{R}} p(s, z-y) z_0(y) dy \right)^2 dz ds.$$

(135) implies that is a C such that

$$\limsup_{s \to 0} s^{1/2} \mathbb{E} \left[\left(\int_{\mathbb{R}} dy p(s, x-y) z_0(y) \right)^2 \right] < \infty. \tag{151}$$

We obtain $E[I_n] \le C''(\frac{t}{4})^{n/2}(n!)^{-1/2}$. Summing in n gives (145). Keeping track of the heat kernels in the last two bounds one obtain (146). □

Now let us go back to (141). If we define $\tilde{f}^n(t) = \sup_{x, s \in [0,t]} s^{1/2} e^{-c|x|} E[|\bar{z}^n(s,x)|^2]$ then we still get

$$f^{n+1}(t) \le C \int_0^t \frac{f^n(s)}{\sqrt{t-s}} ds. \tag{152}$$

It is because we can estimate $\int p^2(t-s, x-y) s^{-1/2} e^{c|y|} dy \le \int p^2(t-s, x-y) s^{-1/2}(e^{cy} + e^{-cy}) dy$ and the latter, by completing the square is equal to $e^{c'(t-s)}(e^{cx} + e^{-cx})(s(t-s))^{-1/2}$. Integrating in s, and noting that t is fixed and finite, so that C could depend on t, gives the bound.

The rest of the argument goes the same way, so we obtain the following existence and uniqueness result:

THEOREM 2.4 (Existence and uniqueness of mild solutions). *Suppose that z_0 satisfies (135). Then there exists a unique progressively measurable $z(t,x)$ satisfying (136) and (138).*

2.5. Martingale problem. The solution $z(t,x)$ of the stochastic heat equation (134) is a random element of $C(\mathbb{R}_+, C(\mathbb{R}))$ where C means the space of continuous functions with the topology of uniform convergence on compact sets. We can also think of it through its distribution, which is a probability measure Q on $C(\mathbb{R}_+, C(\mathbb{R}))$. $C(\mathbb{R}_+, C(\mathbb{R}))$ comes equipped with a natural filtration $\mathscr{F}_t$ which is the σ−field generated by $z(s,x)$, $x \in \mathbb{R}$, $s \le t$. The martingale formulation is an alternative way to identify the probability measure Q by specifying that under it, a rich enough class of processes should be martingales with respect to $\mathscr{F}_t$.

As in the previous section we will need to insist that our probability measure Q satisfies for all $T > 0$ and for some $c < \infty$,

$$\sup_{s\in[0,T]} \sup_{x\in\mathbb{R}} s^{-1/2} e^{-a|x|} E^Q\left[(z(s,x))^2\right] < \infty. \tag{153}$$

DEFINITION 2.5. Such a probability measure Q on $(C(\mathbb{R}^+, C(\mathbb{R})), \mathscr{F})$ is a *solution of the martingale problem for the stochastic heat equation (134) with initial data* z_0 if for all smooth test functions φ on $\mathbb{R}$ with compact support, both

$$N_t(\varphi) = \int_{\mathbb{R}} \varphi(x)z(t,x)dx - \int_{\mathbb{R}} \varphi(x)z_0(x) - \tfrac{1}{2}\int_0^t \int_{\mathbb{R}} \varphi''(x)z(s,x)dxds, \tag{154}$$

and

$$\Lambda_t(\varphi) = N_t(\varphi)^2 - \int_0^t \int_{\mathbb{R}} \varphi^2(x)z^2(s,x)dxds \tag{155}$$

are local martingales under Q.

The martingale problem is extremely useful for studying asymptotic problems for stochastic processes. For example, if we want to prove a discrete KPZ model converges to Hopf-Cole solution of KPZ, we know that we really mean the convergence of the exponentiated version of the former to the solution of the stochastic heat equation. We can do it by writing down approximate versions of (154) and (155) which are martingales in the discrete model. If we can prove tightness, then the limiting process has to solve the martingale problem. The key is therefore the following uniqueness result [**15, 82**].

PROPOSITION 2.6. The martingale problem for the stochastic heat equation has a unique solution which coincides with the distribution of the unique strong solution $z(t,x)$ of (134).

REMARK 2.7. Many readers will be more familiar with the classic martingale problem for finite dimensional stochastic differential equations, which asks that for a nice class of functions f on $\mathbb{R}^d$ (say smooth functions of compact support),

$$M_f(t) := f(x(t)) - \int_0^t Lf(x(s))ds \tag{156}$$

be martingales, where $Lf = \frac{1}{2}\sum_{i,j=1}^d a_{ij}(x)\partial^2_{x_i x_j} f + \sum_{i=1}^d b_i(x)\partial_{x_i} f$. Under reasonable conditions, this produces a weak solution of the stochastic differential equation $dx = b(x)dt + \sigma(x)dB$ on $\mathbb{R}^d$.

Another, older, formulation is simply to ask that this holds for the functions $f(x) = x_i$, $i = 1, \dots, d$ and $f(x) = x_i x_j$, $i, j = 1, \dots, d$ (as local martingales). In fact, we don't even need the quadratic ones if we ask that

$$N_t(i) := x_i(t) - \int_0^t b_i(x(s))ds$$

and

$$\Lambda_t(i,j) := N_t(i)N_t(j) - \int_0^t a_{ij}(x(s))ds$$

are (local) martingales.

Of course, there is no reason to fix a special basis of $\mathbb{R}^d$, and the sum of local martingales is a local martingale. So we may as well take vectors $\varphi \in \mathbb{R}^d$ and ask that $N_t(\varphi) := \varphi \cdot x(t) - \int_0^t \varphi \cdot b(x(s))ds$ and $\Lambda_t(\varphi) = N_t^2(\varphi) - \int_0^t \varphi \cdot a(x(s))\varphi ds$ are martingales. The two equations (154) and (155) are just the infinite dimensional version of this last criterion.

2.6. Chaos representation. Iterating (136) once we obtain

$$\begin{aligned} z(t,x) = &\int_{\mathbb{R}} p(t, x-y)z_0(y)dy + \int_0^t \int_{\mathbb{R}^2} p(t-s, x-y_2)p(s, y_2 - y_1)z_0(y_1) \\ &\times \xi(s, y_1)dy_1 dy_2 ds + \int_{0 \le s_1 < s_2 \le t} \int_{\mathbb{R}^2} p(t - s_2, x - y_2) \\ &\times p(s_2 - s_1, y_2 - y_1)z(s_1, y_1)\xi(s_1, y_1)\xi(s_2, y_2)dy_1 dy_2 ds_1 ds_2. \end{aligned}$$

Iterating this $n-1$ times we obtain,

$$\begin{aligned} z(t,x) = &\sum_{k=0}^{n-1} \int_{0 \le s_1 < \dots < s_k \le t} \int_{\mathbb{R}^{k+1}} p(t - s_k, x - y_k)p(s_k - s_{k-1}, y_k - y_{k-1}) \\ &\cdots p(s_2 - s_1, y_2 - y_1) \times p(s_1, y_1 - y_0)z_0(y_0)\xi(s_1, y_1)\cdots \\ &\xi(s_k, y_k)dy_0 dy_1 \cdots dy_k ds_1 \cdots ds_k + \int_{0 \le s_1 < \dots < s_n \le t} \int_{\mathbb{R}^n} p(t - s_n, x - y_n) \\ &p(s_n - s_{n-1}, y_n - y_{n-1}) \cdots p(s_2 - s_1, y_2 - y_1) \times \\ &\times \ z(s_1, y_1)\xi(s_1, y_1)\cdots\xi(s_n, y_n)dy_1 \cdots dy_n ds_1 \cdots ds_n. \end{aligned}$$

Compute the L^2 norm of the nth term

$$\int_{0\le s_1<\cdots<s_n\le t}\int_{\mathbb{R}^n} p^2(t-s_n,x-y_n)p^2(s_n-s_{n-1},y_n-y_{n-1})$$
$$\cdots p^2(s_2-s_1,y_2-y_1)\times E[z^2(s_1,y_1)]dy_1\cdots dy_n ds_1\cdots ds_n.$$

Use the apriori estimate Lemma 2.3 to see that the nth term goes to 0 in $L^2(P)$ as $n\to\infty$. We obtain the chaos representation

$$z(t,x)=\sum_{k=0}^{\infty}\int_{\Delta_k}\int_{\mathbb{R}^k}\int_{\mathbb{R}} \mathbf{p}_k(\mathbf{s},y_0,\mathbf{y})z_0(y_0)dy_0\xi^{\otimes k}(\mathbf{s},\mathbf{y})d\mathbf{y}d\mathbf{s}. \tag{157}$$

where $\Delta_k=\{0\le s_1<\cdots<s_k\le t\}$ and for such an $\mathbf{s}=(s_1,\ldots,s_k)$

$$\begin{aligned}\mathbf{p}_k(\mathbf{s},y_0,\mathbf{y})&=p(t-s_k,x-y_k)p(s_k-s_{k-1},y_k-y_{k-1})\cdots\\ &\quad p(s_2-s_1,y_2-y_1)p(s_1,y_1-y_0)\end{aligned} \tag{158}$$

are Brownian transition densities to be at y_i at times s_i, $i=1,\ldots,k$ and end up at x at time t, given that the Brownian motion started at y_0 at time 0.

Since

$$\mathbf{p}_k^2(\mathbf{s},\mathbf{y})=\mathbf{p}_k(\mathbf{s},\sqrt{2}\mathbf{y})\prod_{j=1}^{k}\frac{1}{\sqrt{2\pi(s_j-s_{j-1})}},$$

we have

$$\int_{\Delta_k}\int_{\mathbb{R}^k}\mathbf{p}_k^2(\mathbf{s},\mathbf{y})\,d\mathbf{y}\,d\mathbf{s}=(4\pi)^{-\frac{k}{2}}\int_{\Delta_k}\prod_{j=1}^{k}\frac{1}{\sqrt{s_j-s_{j-1}}}d\mathbf{s}=\frac{1}{2^k\Gamma\left(\frac{k}{2}+1\right)}. \tag{159}$$

The second equality comes from recognizing that the integrand is the density of a Dirichlet distribution. So (157) is easily seen to be a valid chaos series.

The chaos series represents an alternative way to prove discrete models converge to the stochastic heat equation. For example, the directed random polymer partition functions have the form of discrete versions of the chaos series.

To illustrate another use of the chaos series, we have the following small time asymptotics for the solution of the stochastic heat equation with Dirac initial data, which is readily checked by computing the mean square of the $k\ge 2$ terms in the chaos series (157) are of smaller order, so that the first (deterministic) term and the second (linear) term dominate.

PROPOSITION 2.8. *Let $z(t,x)$ be the solution of (134) with $z(0,x)=\delta_0(x)$. As $t\searrow 0$,*

$$z(t,x)=p(t,x)+t^{1/4}\zeta(t,x)+\mathcal{O}(t^{1/2}) \tag{160}$$

where the process $\zeta(t,x)=t^{-1/4}\int_0^t\int_{\mathbb{R}}p(t-s,x-y)p(s,y)\xi(s,y)dyds$ is Gaussian mean zero, with finite limiting covariance as $t\searrow 0$.

So for small time, the heat part of the equation dominates. Note that on the physical side this is related to the picture of the weakly asymmetric scaling being critical between the KPZ and the Edwards-Wilkinsen (EW) universality classes. The relation between t and β in the continuum random polymer is $t \sim \beta^4$, as can be seen by a naive Brownian scaling. So $t \to 0$ is the same as $\beta \to 0$, and as the asymmetry goes away, one finds oneself in the EW class. Another way to think about it is to put a small parameter in front of the non-linearity in KPZ and send it to zero, to see the Gaussian fluctuations.

2.7. Regularity.

THEOREM 2.9. *Let $z(t,x)$ be the solution to (134) with $z_0(x) \geq 0$ satisfying (135). For any $\alpha < 1/2$ and $\beta < 1/4$, and any $\delta > 0$, $T < \infty$,*

$$\lim_{\lambda\to\infty} P\left(\sup_{\delta \leq s < t \leq T,\ |x|,|y| \leq B,} \frac{|z(t,x) - z(s,y|}{|t-s|^\beta + |x-y|^\alpha} \geq \lambda \right) = 0 \tag{161}$$

This can be proved with the methods of Section 2.13. Since it is almost the same, we will not repeat it here. There are many references, e.g. [**119**], [**45**].

2.8. Comparison. We are interested in studying stochastic partial differential equations of the type

$$\begin{aligned} &\partial_t z = \tfrac{1}{2}\partial_x^2 z + f(z) + \sigma(z)\xi, \qquad t>0,\ x\in\mathbb{R}, \\ &z(0,x) = z_0(x). \end{aligned} \tag{162}$$

Of course, this means the Duhamel version

$$z(t,x) = \int_{\mathbb{R}} p(t,x-y)z_0(y)dy + \int_0^t \int_{\mathbb{R}} p(t-s,x-y)f(z(s,y))dyds \tag{163}$$

$$+ \int_0^t \int_{\mathbb{R}} p(t-s,x-y)\sigma(z(s,y))\xi(s,y)dyds. \tag{164}$$

If f and σ are Lipschitz, then the existence and uniqueness theorem still applies (see [**45**]).

There are not many general methods for studying stochastic partial differential equations. One very disturbing thing to keep in mind is that if z satisfies (162) and Φ is non-linear in the sense that $\Phi''(z) \neq 0$ then $\partial_t \Phi(z)$ does *not* satisfy a stochastic partial differential equation. The reason is that unlike the case of finite dimensional stochastic differential equations, the Itô correction is always infinite. One way out is to consider *coloured noises* where the correlations $\langle \xi(t,x), \xi(s,y)\rangle = \phi(y-x)\delta(t-s)$ are more reasonable in space. This is much the same thing as discretizing the equation,

$$\partial_t z_{\frac{i}{n}} = \tfrac{1}{2}[z_{\frac{i+1}{n}} - 2z_{\frac{i}{n}} z_{\frac{i-1}{n}}] + f(z_{\frac{i}{n}}) + n^{1/2}\sigma(z_{\frac{i}{n}})dB_{\frac{i}{n}}. \tag{165}$$

One of the few general facts that are available is the *comparison principle.*

PROPOSITION 2.10. *Let $z_1(t,x), z_2(t,x)$ be two solutions of (162) with $f_1(z) \le f_2(z)$ and $\sigma_1(z) = \sigma_2(z)$ Lipschitz functions, both using the same white noise ξ and with initial data*

$$z_1(0,x) \le z_2(0,x), \qquad x \in \mathbb{R}$$

Then, with probability one, for all $t \ge 0$,

$$z_1(t,x) \le z_2(t,x), \qquad x \in \mathbb{R}.$$

The proof is via approximation by discretizations (165), for which the comparison is almost obvious. Note that the Lipschitz assumption is only used in the uniqueness. In its absence one still has versions for which the comparison is true, which can be sufficient to prove distributional statements about the solutions (see **[93]** for an example.) In the case of the stochastic heat equation (134), we can also prove Proposition 2.10 by approximation with asymmetric exclusion (see Section 201). The comparison at the discrete level follows from the basic coupling for exclusion.

2.9. Positivity.

THEOREM 2.11 (Mueller 91 **[92]**). *Let $z(t,x)$ be the solution to (134) with $z_0(x) \ge 0$ and $\int z_0(x)dx > 0$. Then for all $t > 0$, $z(t,x) > 0$ for all $x \in \mathbb{R}$ with probability one.*

REMARK 2.12. It is also true that $z(t,x) > 0$ for all $x \in \mathbb{R}$ and $t > 0$ with probability one, but we only give the proof of the weaker statement.

REMARK 2.13. We could also start with a non-trivial positive measure. In a small time the heat part of the semigroup sends us in to the situation above.

REMARK 2.14. If the theorem seems obvious, note that it is only *just* true. If $\frac{1}{2} < \gamma < 1$ then if z_0 has compact support, the solution of $\partial_t z = \frac{1}{2}\partial_x^2 z + \xi z^\gamma$ does as well **[74]**. When z is small, it is a battle between the heat equation trying to spread the stuff out and the random term killing the stuff. If $\gamma < 1$ then the random term is too large and it wins. If $\gamma \ge 1$ the random term is too small and the heat part wins.

Let

$$N(t,x) = \int_0^t \int_{\mathbb{R}} p(t-s, x-y) f(s,y) \xi(s,y) dy ds. \tag{166}$$

If $f(s,y) = z(s,y)$ then this is the noise term in (136). We need to show that it doesn't win against the heat equation smoothing, so we need an estimate of how big it is. It is a random function of t and x (a random field) and we need to control the maximum. If the random f were not in the integrand then we just have a mean zero Gaussian field. We compute the covariance, and knowing this, the maximum can be controlled in terms of a typical point. The main point of the following large deviations estimate is that roughly the same thing is true for (166).

LEMMA 2.15. *Suppose that f is progressively measurable with $|f(s,y)| \le K$, then, for $r<\infty$ there exists $0<C<\infty$ such that*

$$P(\sup_{\substack{0\le t\le T\\ |x|\le r}} |N(t,x)| \ge \lambda) \le C\exp\left\{-C\lambda^2 T^{-1/2}K^{-2}\right\}. \tag{167}$$

A proof of the lemma can be found on page 30 of [**45**]. We sketch the original proof of Theorem 2.11 due to Mueller [**92**]. It would be nice to have alternative arguments.

PROOF OF THEOREM 2.11. We have $z_0(x) \ge \gamma \mathbf{1}_{[b-a,b+a]}(x)$ for some $a>0$, b and $\gamma>0$. If we shift the noise by b along with the initial condition the solution just gets shifted as well. So without loss of generality, we can assume that $b=0$. Furthermore, if $z(t,x)$ is a solution of the stochastic heat equation (134) then so is $\gamma^{-1}z(t,x)$. So we can also assume $\gamma=1$. By the comparison principle, Proposition 2.10, we therefore have the following problem: *Let $t>0$. Under the assumption that $z(0,x)=\mathbf{1}_{[-a,a]}(x)$, show that* $P(z(t,x)>0$ for $|x|\le M) > 1-\delta$ *for all $M,\delta>0$.* To prove this, let

$$\mathscr{A}_k = \left\{z(\tfrac{tk}{\ell},x) \ge \tfrac{1}{8^k}\mathbf{1}_{[-a-Mk\ell^{-1},a+Mk\ell^{-1}]}(x)\right\}. \tag{168}$$

Suppose we could show that

$$P(\mathscr{A}_{k+1}^C \mid \mathscr{A}_1\cap\cdots\cap\mathscr{A}_k) < \tfrac{\delta}{\ell}. \tag{169}$$

Then we would have $P(\mathscr{A}_n^C) \le \sum_{k=0}^{n-1} P(\mathscr{A}_{k+1}^C \mid \mathscr{A}_1\cap\cdots\cap\mathscr{A}_k) < \delta$ as desired.

So we concentrate on (169). First of all note that by the Markov property, it is the same thing as proving $P(\mathscr{A}_{k+1}^C \mid \mathscr{A}_k) < \frac{\delta}{\ell}$. By the comparison principle Proposition 2.10, and the linearity of the stochastic heat equation (134) in the initial data, it is enough to show that (*) *there is an $\ell>0$ so that if we start with $z(0,x)=\mathbf{1}_{[-a-\frac{Mk}{\ell},a+\frac{Mk}{\ell}]}(x)$, then $z(\frac{t}{\ell},x) \ge \frac{1}{8}\mathbf{1}_{[-a-M(k+1)\ell^{-1},a+M(k+1)\ell^{-1}]}(x)$ with probability at least $1-\delta/\ell$.*

To show (*), note that

$$z(\tfrac{t}{\ell},x) = \int_{a-\frac{Mk}{\ell}}^{a+\frac{Mk}{\ell}]} p(\tfrac{t}{\ell},x-y)dy - \int_0^{\frac{t}{\ell}}\int_{\mathbb{R}} p(\tfrac{t}{\ell}-s,x-y)z(s,y)\xi(s,y)dyds. \tag{170}$$

The key point is now that for small times, the first, deterministic, term dominates. We are interested in a lower bound for it for x on $[-a-M(k+1)\ell^{-1}, a+M(k+1)\ell^{-1}]$. By inspection, the minimum is at $x=a+M(k+1)\ell^{-1}$, where by normalizing the Gaussian integral,

$$\begin{aligned} &\min_{|x|\le a+M(k+1)\ell^{-1}} \int_{a-\frac{Mk}{\ell}}^{a+\frac{Mk}{\ell}} p(\tfrac{t}{\ell},x-y)dy \\ &= \int_{Mt^{-1/2}\ell^{-1/2}}^{2at^{-1/2}\ell^{1/2}+M(2k+1)t^{-1/2}\ell^{-1/2}} p(1,x)dx > \tfrac14 \end{aligned} \tag{171}$$

for sufficiently large ℓ. Hence what we need to show is that for sufficiently large ℓ, with probability at least $1 - \delta/\ell$,

$$\sup_{|x| \le a + M(k+1)\ell^{-1}} \left| \int_0^{\frac{t}{\ell}} \int_{\mathbb{R}} p(\tfrac{t}{\ell} - s, x - y) z(s, y) \xi(s, y) dy ds \right| < \tfrac{1}{8}. \tag{172}$$

The proof will be based on the large deviation estimate Lemma 2.15. But first let's see roughly why it is true. One believes that the supremum of such objects behaves, up to constants, like a typical value. So we drop the sup and compute the variance

$$E\left[\left(\int_0^{\frac{t}{\ell}} \int_{\mathbb{R}} p(\tfrac{t}{\ell} - s, x - y) z(s, y) \xi(s, y) dy ds\right)^2\right]$$

$$= \int_0^{\frac{t}{\ell}} \int_{\mathbb{R}} p^2(\tfrac{t}{\ell} - s, x - y) E[z^2(s, y)] dy ds. \tag{173}$$

Recall that $E[z^2(s, y)] \le C u^2(s, y)$ where u solves the deterministic heat equation with the same initial data. So the variance is bounded by

$$\int_0^{\frac{t}{\ell}} \int_{\mathbb{R}} p^2(\tfrac{t}{\ell} - s, x - y) \left(\int_{|b| \le a + Mk\ell^{-1}} p(s, y - b) db\right)^2 dy ds \to 0 \tag{174}$$

as $\ell \to \infty$. In other words, we should be able to choose an ℓ large enough that (172) holds.

To make it rigorous, we can assume without loss of generality that there is a $K < \infty$ such that $z(t, x) < K$ throughout, for by the regularity theorem, $z(t, x)$ is continuous and bounded, so $P(\sup_{\substack{0 \le s \le t \\ |x| \le r}} z(t, x) > K) \to 0$, so we can choose K large enough that this probability is less than $\delta/2$. Then, by Lemma 2.15, the probability of the complement of (172) is bounded by $C \exp\left\{-C\ell^{1/2} t^{-1/2} K^{-2}\right\}$ which is less than $\delta/2$ for sufficiently large ℓ. □

REMARK 2.16. It is worth noting that the proof after (170), comparing the deterministic and stochastic parts, works also for $\partial_t z = \frac{1}{2}\partial_x^2 z + z^\gamma \xi$ when $\gamma < 1$, although we know that in that case, the stochastic part wins and the solution has compact support. The reason the proof does not work is that at each step k, in the reduction to (*), in the $\gamma < 1$ case one ends up increasing the noise. If $\gamma > 1$, the noise is reduced at each iteration. This is why the linear case is critical for the argument.

2.9.1. *Regularity of the one-point density.* Let $z(t, x)$ be the solution of the stochastic heat equation. We will be studying $F(s) = P(z(t, x) \le s)$. The density (if it exists) is $f(s) = F'(s)$. Using Malliavin calculus it can be shown[6] **[98]**,**[91]** that $f(s)$ is a smooth function. The key ingredient is the fact that $z(t, x) > 0$.

[6]The results are for the stochastic heat equation on a finite interval with Dirichlet boundary conditions

2.10. The continuum random polymer. Fix a space-time white noise $\xi(t,x)$. The (point-to-point) continuum random polymer is the probability measure $P^{\xi}_{0,0,T,x}$ on continuous functions $x(t)$ on $[0,T]$ with $x(0)=0$ and $x(T)=x$ and formal density

$$e^{\int_0^T \xi(t,x(t))dt} \tag{175}$$

with respect to Brownian motion. Unfortunately the density does not make sense. Once again, it needs a type of infinite renormalization. This can be achieved in several ways. We could smooth out the noise, construct the measure with the smoothed out noise, and then obtain our desired measure as a limit as the smoothing is removed. We will give an alternate construction which is direct and stresses the Markov property.

Let $z(s,x,t,y)$ denote the solution of the stochastic heat equation after time s starting with a delta function at x,

$$\begin{aligned} \partial_t z &= \tfrac{1}{2}\partial_y^2 z + \xi z, \qquad t > s,\ y \in \mathbb{R}, \\ z(s,x,s,y) &= \delta_x(y). \end{aligned} \tag{176}$$

It is important that they are all using the same noise ξ.

$P^{\xi}_{0,0,T,x}$ is defined to be the probability measure on continuous functions $x(t)$ on $[0,T]$ with $x(0)=0$ and $x(T)=x$ and finite dimensional distributions

(177)

$$P^{\xi}_{0,0,T,x}(x(t_1)\in dx_1,\dots,x(t_n)\in dx_n) = \frac{z(0,0,t_1,x_1)z(t_1,x_1,t_2,x_2)\cdots z(t_{n-1},x_{n-1},t_n,x_n)z(t_n,x_n,T,x)}{z(0,0,T,x)} dx_1\cdots dx_n$$

for $0 < t_1 < t_2 < \cdots < t_n < T$.

One can check these are a consistent family of finite dimensional distributions. It is basically because of the Chapman-Kolmogorov equation

$$\int_{\mathbb{R}} z(s,x,\tau,u)z(\tau,u,t,y)du = z(s,x,t,y) \tag{178}$$

which is a consequence of the linearity and uniqueness of solutions of the stochastic heat equation. So there is a probability measure on the product space. One can then show using essentially the Kolmogorov criteria, that the paths are Hölder of any exponent less than $1/2$, just like Brownian motion. In particular, they are continuous. Using (178) the resulting process is Markovian. The quadratic variation can be computed and it is the same as Brownian motion. However, for almost every realization of the noise ξ, $P^{\xi}_{0,0,T,x}$ is singular with respect to the Brownian bridge $P^{0}_{0,0,T,x}$ (see [**1**]for proofs).

One can also define the point-to-line continuum random polymer $P^{\xi}_{0,0}$ on continuous functions $x(t)$ on $[0,T]$ with $x(0)=0$. Its finite dimensional distributions are

(179)

$$P^{\xi}_{0,0}(x(t_1)\in dx_1,\dots,x(t_n)\in dx_n, x(t)\in dx) = \frac{z(0,0,t_1,x_1)z(t_1,x_1,t_2,x_2)\cdots z(t_{n-1},x_{n-1},t_n,x_n)z(t_n,x_n,T,x)}{\int_{\mathbb{R}} z(0,0,T,x)dx} dx_1\cdots dx_n dx$$

The paths $x(t)$ look like Brownian motions under $P^{\xi}_{0,0}$ on a small scale, but on a large scale (i.e. for large T) they are expected to have the scaling $x(T)\sim T^{2/3}$.

The continuum random polymer is the limit of diffusively rescaled polymer paths in the intermediate scaling regime as long as the random environment satisfies $E[e^{\lambda\omega}]<\infty$ for sufficiently small $|\lambda|$ [**1**] (note however, that it is expected to be true as long as $E[\omega_-^6]<\infty$.)

2.11. Universality of KPZ and the continuum random polymer: The weakly asymmetric limit. There are two other choices of scaling in (15). 1. If one takes $b=1/2$ and $z=2$, then (13) turns

$$\partial_t h = -\tfrac{1}{2}\epsilon^{1/2}(\partial_x h)^2 + \tfrac{1}{2}\partial_x^2 h + \xi \tag{180}$$

into

$$\partial_t h_\epsilon = -\tfrac{1}{2}(\partial_x h_\epsilon)^2 + \tfrac{1}{2}\partial_x^2 h_\epsilon + \xi. \tag{181}$$

This is the idea behind the weakly asymmetric limit in which the KPZ equation can be obtained as a limit of asymmetric exclusion. 2. If one takes $b=0$ and $z=2$ then (13) turns

$$\partial_t h = -\tfrac{1}{2}(\partial_x h)^2 + \tfrac{1}{2}\partial_x^2 h + \epsilon^{1/2}\xi \tag{182}$$

into

$$\partial_t h_\epsilon = -\tfrac{1}{2}(\partial_x h_\epsilon)^2 + \tfrac{1}{2}\partial_x^2 h_\epsilon + \xi. \tag{183}$$

This is the idea behind the intermediate disorder limit in which the KPZ equation can be obtained as a limit of directed random polymers with small β.

If a model in the KPZ class has an adjustable parameter, then a diffusive scaling of time and space combined with this weak asymmetry/disorder can lead to KPZ. The KPZ equation is thus a fixed point of its own small universality class. This is often called the universality of the KPZ equation, which should not be confused with the KPZ universality, in which the fixed point is highly trivial and poorly understood, with Airy processes as its fixed time marginals.

2.12. Intermediate disorder regime for directed random polymers. We want to study the intermediate disorder limit of the free energy given by (58) with $x = 0$. The weakly asymmetric limit described in the previous sections tells us to look at space scales ϵ^{-1}, time scales ϵ^{-2} and β of order $\epsilon^{1/2}$. We define

$$z_\epsilon(t,x) = c_\epsilon(t) Z_{\beta\epsilon^{1/2}}(0, \lfloor \epsilon^{-2} t \rfloor, \lfloor \epsilon^{-1} x \rfloor) \tag{184}$$

Note that β has changed its meaning. The constant c_ϵ depends on the model and consists of two pieces: There is a factor ϵ^{-1} so that the initial data for z_ϵ looks like a Dirac delta function at 0. There is also a factor $(1 - E[e^{-\beta\epsilon^{1/2}\xi} - 1])^{\lfloor \epsilon^{-2} t \rfloor}$ which compensates for the deterministic growth term which is hiding in the equation. So

$$c_\epsilon(t) = \epsilon^{-1}(1 - E[e^{-\beta\epsilon^{1/2}\xi} - 1])^{\lfloor \epsilon^{-2} t \rfloor}. \tag{185}$$

Note that because of parity, the process $\tilde{z}_\epsilon(t,x)$ lives on the *even*[7] sites of the lattice $\epsilon^2\mathbb{Z} \times \epsilon\mathbb{Z}$. Now let $p_\epsilon(t,x)$ be the solution of the discrete heat equation

$$p_\epsilon(t+\epsilon^2, x) = \tfrac{1}{2}(p_\epsilon(t, x+\epsilon) + p_\epsilon(t, x-\epsilon)) \tag{186}$$

with initial data $\epsilon^{-1}\mathbf{1}(x=0)$. Note that we have implicitly set everything to 0 on the odd sites of the lattice. After some computation we obtain from (58) that

$$z_\epsilon(t,x) = p_\epsilon(t,x) + 2^{-1/2}\beta N^{-1/2} \sum_{s\in\epsilon^2\mathbb{Z}\cap[0,t], x\in\epsilon\mathbb{Z}} p_\epsilon(t-s, x-y)\hat{z}_\epsilon(s-\epsilon^2, y)\xi_\epsilon(s,y) \tag{187}$$

where $\hat{z}_\epsilon(s,y) = \frac{1}{2}(z_\epsilon(s, y+\epsilon) + z_\epsilon(s, y-\epsilon))$,

$$\xi_\epsilon = \beta^{-1}\epsilon^{-1/2}(e^{-\beta\epsilon^{1/2}\xi} - E[e^{-\beta\epsilon^{1/2}\xi}]), \tag{188}$$

and $N = \epsilon^{-3}/2$ is the number of summands. This equation is clearly a discrete version of the mild form (136) of the stochastic heat equation. Then we can repeat the argument of Lemma 2.3 and we will show in Section 2.13 how to prove that the z_ϵ are tight as two parameter stochastic processes. Once we have tightness, we have weak convergence, and the limit will satisfy the mild form (136) of the stochastic heat equation, thus identifying it as the unique solution. Note the convergence is a little subtle. While it is true that $p_\epsilon(t,x) \to p(t,x)$ weakly, it is *not* true strongly. But it is just because of the parity problem: $p_\epsilon(t,x)$ takes twice the normal value on sites with even parity and vanishes on sites with odd parity. The same thing happens for $z_\epsilon(t,x)$. Multiplied, this gives a factor of 4 on the last term in (187), as

[7] In fact, the $\xi(i,j)$ where i and j do not have the same parity are not used in the directed polymer problem.

opposed to the first two terms which have a factor of 2. This results in the limiting equation

$$(189)\quad z_{\sqrt{2}\beta}(t,x) = p(t,x) + \sqrt{2}\beta \int_0^t \int_{\mathbb{R}} p(t-s,x-y) z_{\sqrt{2}\beta}(s,y)\xi(s,y)dyds$$

or

$$(190)\qquad \partial_t z_{\sqrt{2}\beta} = \tfrac{1}{2}\partial_x^2 z_{\sqrt{2}\beta} + \sqrt{2}\beta\xi z_{\sqrt{2}\beta}, \qquad z_{\sqrt{2}\beta}(0,x) = \delta_0(x).$$

This was originally done by expanding the discrete partition function as a series of discrete Wiener-Itô chaose series and taking the limit term by term [**2**].

Before we turn to the tightness, let us make a few remarks and state the general result.

REMARK 2.17. The above argument shows that the rescaled unconditioned point-to-point partition function converges to the stochastic heat equation. By unconditioned, we just mean that we take the expectation on the set of paths which end up at a point, instead of conditioning them to end up at that point. To get the conditioning, one just divides by the probability to end at that point, which includes a term of order ϵ, cancelling the ϵ^{-1} in (185). To get the point-to-line partition function asymptotics also requires at most very slight tweaks of the above proof.

REMARK 2.18. It would appear that the argument depends on the existence of exponential moments of the random field ξ. In fact, one only needs the finiteness of a few moments. Without the intermediate scaling, the free energy is predicted to have the Tracy-Widom asymptotics as soon as $E[\xi_-^5] < \infty$ where $\xi_- = \max(-\omega, 0)$ [**18**]. A similar reasoning leads to the prediction that for the intermediate scaling, $E[\xi_-^6] < \infty$ is enough. Let us explain roughly the argument when $E[\xi_-^8] < \infty$, which is simpler. In the argument above, we are making an expansion of the exponetial in (188) so it goes wrong when $\xi \ll -\epsilon^{-1/2}$. By Chebyshev's inequality,

$$(191)\qquad P(\xi < -\epsilon^{-1/2}) \le \epsilon^4 E[\xi_-^8 \mathbf{1}_{\xi_- < \epsilon^{-1/2}}] = o(\epsilon^4).$$

There are only order ϵ^{-4} sites in our lattice that the random walk path can possibly visit. Hence the probability that even one site has $\xi < -\epsilon^{-1/2}$ goes to zero with ϵ. So we can work with the cutoff variables without affecting the result. To see that $E[\xi_-^6] < \infty$ should really be enough, just note that the random walk paths really only visit ϵ^{-3} sites. In fact, one expects the condition $\epsilon^{-3}P(\xi < -\epsilon^{-1/2}) \to 0$ as $\epsilon \to 0$ to be necessary and sufficient.

To state the main result, we extend the field $z_\epsilon(t,x)$ of (184) which in principle lives on $\epsilon^2\mathbb{Z}_+ \times \epsilon\mathbb{Z}$, to $\mathbb{R}_+ \times \mathbb{R}$. It doesn't really matter how we do this, but we could first make the paths continuous in x by drawing little straight lines between $z_\epsilon(t,x)$ and $z_\epsilon(t,x+\epsilon)$. Then we could draw straight lines from $z_\epsilon(t,x)$ to $z_\epsilon(t+\epsilon^2,x)$. The result is a continuous function on

$\mathbb{R}_+ \times \mathbb{R}$. Let's call its distribution P_ϵ. So for each $\epsilon > 0$ we have a probability measure on $C([0,\infty), C(\mathbb{R}))$ where C means continuous functions with the topology of uniform convergence on compact sets. The main result is

THEOREM 2.19 ([**2**]). *Suppose that $E[\xi_-^8] < \infty$. The measures P_ϵ are tight and have a unique weak limit P which is the distribution of the solution of the stochastic heat equation (189) with initial condition $z(0,x) = \delta_0(x)$.*

We have sketched above the argument to identify the limit. Below we sketch the argument for the tightness. First we note an important corollary. Recall the key open problem for directed polymers in one dimension is to show that the free energy has the asymptotics

$$\log E_x\left[e^{-\beta\sum_{i=0}^N \xi(i,x_i)}\right] \sim cN + N^{1/3}\zeta \tag{192}$$

where the fluctuations ζ have the GUE Tracy-Widom distribution. It is supposed to be true for any distribution for the i.i.d. ξ satisfying $E[\xi_-^5] < \infty$, though not a single example is known[8]. Combining Theorem 2.19 and Theorem 1.5 we have

COROLLARY 2.20 (Weak universality). *Suppose that $E[\xi_-^8] < \infty$ and without loss of generality assume $E[\xi] = 0$. Let $\kappa_2 = E[\xi^2]$, $\kappa_3 = E[\xi^3]$, $\kappa_4 = E[\xi^4 - 3\xi^2]$ be the next three cumulants. Let E_x denote the expectation over the random walk conditioned to start at 0 and end at x after n steps. Then*

$$\lim_{\beta\to\infty}\lim_{N\to\infty} \frac{\log E_x\left[e^{-\frac{\beta}{N^{1/4}}\sum_{i=0}^N \xi(i,x_i)}\right] - \frac{\kappa_2}{2}\beta^2 N^{1/2} + \frac{\kappa_3}{3!}\beta^3 N^{1/4} - \frac{\kappa_4}{4!}\beta^4}{\beta^{4/3}}$$

$$\stackrel{\text{dist}}{=} F_{GUE} \tag{193}$$

As discussed above, the we expect it to hold when $\epsilon^{-3}P(\xi < -\epsilon^{-1/2}) \to 0$.

2.13. Tightness of the approximating partition functions. We just give a sketch of the tightness. It comes down to obtaining some modulus of continuity estimates which hold uniformly in $\epsilon > 0$. Start with the equation for (185) which we can write as

$$z_\epsilon(t,x) = \epsilon \sum_{y\in\epsilon\mathbb{Z}} p_\epsilon(t-\delta, x-y) z_\epsilon(\delta,y) + \epsilon^2$$
$$\sum_{s\in[\delta,t]\cap\epsilon^2\mathbb{Z}} \epsilon \sum_{y\in\epsilon\mathbb{Z}} p_\epsilon(t-s,x-y) z_\epsilon(s,y)\xi_\epsilon(s,y) \tag{194}$$

for any $\delta \in [0,t)$. Here $p_\epsilon(t,x)$ are the discrete heat kernels, i.e. the transition densities of the underlying random walks, normalized so that as ϵ tends

[8] During the final preparation of this manuscript the result was proved for the special case of ξ with the log gamma distribution [**22**].

to zero, $p_\epsilon(t,x)$ tends to the standard Brownian transition probability density $p(t,x) = \frac{1}{\sqrt{2\pi t}} e^{-x^2/2t}$. In exactly the same way as in (146) we get the apriori bound

$$E\left[z_\epsilon^2(t,y)\right] \le C p_\epsilon^2(t,y). \tag{195}$$

By Burkholder's inequality applied to the last term in (194) we also have for $M > 2$,

$$E\left[z_\epsilon^M(s,y)\right] \le C_M p_\epsilon^M(s,y). \tag{196}$$

Note that for the first term of (194) the bound is just standard for the (discrete) heat equation. From (195) we also have good modulus of continuity estimates for the first term on the right side of (194). Now we try to develop them for the second term which we write as

$$U_{\epsilon,\delta}(x,t) = \int_\delta^t \int p_\epsilon(t-s,x-y) z_\epsilon(s,y) \xi_\epsilon(s,y) \tag{197}$$

introducing the integral notation for the sums to avoid huge expressions. At any rate, if we extend these functions from the lattice to $\mathbb{R}_+ \times \mathbb{R}$, the sums really are replaced by integrals. By Burkholder's inequality, we can estimate $E[|U_\epsilon(x+\gamma,t) - U_\epsilon(x,t)|^M]$ by

$$C_M E\left[\left(\int_\delta^t \int (p_\epsilon(t-s,x+\gamma-y) - p_\epsilon(t-s,x-y))^2 z_\epsilon^2(s,y)\right)^{M/2}\right] \tag{198}$$

Apply Hölder with $p = M/2, q = M/(M-2)$ to bound this by

$$\begin{aligned} & C'_M E\left[\int_\delta^t \int z_\epsilon^M(s,y)\right] \\ & \times \left[\int_\delta^t \int (p_n(t-s,x+\gamma-y) - p_n(t-s,x-y))^{\frac{2M}{M-2}}\right]^{\frac{M-2}{2}} \end{aligned}$$

From (196) we can bound on the first term independent of ϵ (it does depend on t, δ). And one can check that there is a C also depending only on $t, \epsilon > 0$ such that

$$\left[\int_\delta^t \int (p_\epsilon(t-s,x+\gamma-y) - p_\epsilon(t-s,x-y))^{\frac{2M}{M-2}}\right]^{\frac{M-2}{2}} \le C\gamma^{\frac{M}{2}-1}.$$

Now we try to do the same thing for $E[|U_\epsilon(x,t+h) - U_\epsilon(x,t)|^M]^{1/M}$. We estimate it by constant multiples of two terms

$$E\left[\left|\int_\delta^t \int (p_\epsilon(t+h-s,x-y) - p_\epsilon(t-s,x-y))^2 z_\epsilon^2(s,y)\right|^{\frac{M}{2}}\right]^{\frac{1}{M}} \tag{199}$$

and

$$E\left[\left|\int_t^{t+h} \int p_\epsilon^2(t+h-s,x-y) z_\epsilon^2(s,y)\right|^{\frac{M}{2}}\right]^{\frac{1}{M}} \tag{200}$$

The first one we estimate by Hölder with $p = M/2, q = M/(M-2)$,

$$E\left[\int_\delta^t \int z_\epsilon^M(s,y)\right]^{\frac{1}{M}}$$

$$\times \left(\int_\delta^t \int |p_\epsilon(t+h-s, x-y) - p_\epsilon(t-s, x-y)|^{\frac{2M}{M-2}}\right)^{\frac{M-2}{2M}}$$

The apriori bound (195) controls the expectation. The second term is bounded in ϵ when $M > 8$, in which case it can be computed and is less than $C_M h^{\frac{1}{4}-\frac{2}{M}}$. (200) we estimate by Hölder again

$$E\left[\int_t^{t+h} \int z_\epsilon^M(s,y)\right]^{\frac{1}{M}} \left(\int_\delta^t \int p_\epsilon^2(t-s, x-y)\right)^{\frac{M-2}{2M}} \tag{201}$$

From the apriori bound this is again less than $C_M h^{\frac{1}{4}-\frac{2}{M}}$.

We obtain

LEMMA 2.21. *For each even $M > 8$, and each $\delta > 0$, there is a $C_M < \infty$ such that for $t, t+h \geq \delta$,*

$$E[|z_n(x+\gamma, t+h) - z_n(x,t)|^M]^{1/M} \leq C_M \max(\gamma, h)^{\frac{1}{4}-\frac{2}{M}}. \tag{202}$$

Note that if we were more careful we could improve the modulus of continuity to Hölder 1/2− in space, but for tightness we do not need an optimal result.

Now we use the inequality of Garsia [**64**] that

$$|f(x) - f(y)| \leq 8 \int_0^{|x-y|} \Psi^{-1}(B/u^{2d})dp(u) \tag{203}$$

for all functions f continuous in a unit cube $I \subset \mathbf{R}^d$ that satisfy the inequality

$$\int_I \int_I \Psi(f(x) - f(y)/p(d^{-1/2}|x-y|))\, dxdy \leq B$$

for all hypercubes $I \subset I_0$, where (i) $e(I)$ denotes the volume of I, (ii) Ψ is non-constant even convex function and (iii) p is a continuous even function that satisfies the condition $\lim_{n\to 0} p(u) = 0$.

We are working in $d = 2$ (space+time). Choosing $\Psi(x) = x^M$, $M > 6$ and $p(x) = x^{\gamma/M}$ we have from Lemma 2.21.

$$E\left[\int_{t,s\in[\delta,T],x,y\in\mathbb{R}} \Psi\left(\frac{|z_n(x,t) - z_n(y,s)|}{p(2^{-1/2}\sqrt{(t-s)^2 + (x-y)^2})}\right)\right] \leq C_M \tag{204}$$

Here we have extended $z_n(t,x)$ as a continuous function to $[0,T] \times \mathbb{R}$.

Since $\int_0^h \Psi^{-1}(B/u^{2d})dp(u) = C_{M,\gamma} B^{1/M} h^{\frac{\gamma-4}{M}}$ with a finite $C_{M,\gamma}$ for $\gamma > 4$, we conclude that if $\mathcal{H}_{[\delta,T]\times\mathbb{R}}(\alpha, K)$ denotes the set of functions $z(t,x)$ on $[\delta,T] \times \mathbb{R}$ with

$$|z(t,x) - z(s,y)| \leq K|(t-s)^2 + (x-y)^2|^{\alpha/2},$$

then

LEMMA 2.22. *If P_n denotes the distribution of $z_n(t,x)$ then for any $\delta > 0$ and $\alpha < 1/4$,*

$$\limsup_{K\to\infty}\limsup_{n\to\infty} P_n(\mathcal{H}_{[\epsilon,T]\times\mathbb{R}}(\alpha,K)) = 1. \tag{205}$$

In particular, since $\mathcal{H}_{[\epsilon,T]\times\mathbb{R}}(\alpha,K)$ are compact sets of $C([\delta,T]\times\mathbb{R})$, the P_n are tight.

2.14. Asymmetric exclusion. The asymmetric simple exclusion process (ASEP) with parameters $p,q\geq 0$, $p+q=1$, $p\neq q$ is a continuous time Markov process on $S=\{0,1\}^{\mathbb{Z}}$, the 1's being thought of as particles and the 0's as holes. Each particle waits a random exponent mean one amount of time and then attempts a jump, one site to the right with probability p and one site to the left with probability q. However, the jump is only performed if there is no particle at the target site. Otherwise, nothing happens and the particle waits another exponential time. All particles are doing this independently of each other.

The generator of ASEP acts on local functions (functions which depend on only finitely many coordinates) by

$$Lf(\eta) = \sum_x \{p\eta(x)(1-\eta(x+1)) + q(1-\eta(x))(\eta(x+1))\}\,(f(\eta^{x,x+1})-f(\eta)) \tag{206}$$

where $\eta^{x,x+1}\in\{0,1\}^{\mathbb{Z}}$ is obtained from η by switching the occupation variables at x and $x+1$. The corresponding Markov semigroup acts on the Banach space $C(S)$ of continuous functions on S with sup norm $\|f\| = \sup_{\eta\in S}|f(\eta)|$ by

$$P_t f(\eta) = E[f(\eta(t)) \mid \eta(0)=\eta]. \tag{207}$$

The domain of the generator L is given by

$$\mathcal{D} = \left\{ f\in C(S) : Lf = \lim_{t\searrow 0} t^{-1}(P_t f - f) \text{ exists}\right\}.$$

The problem with such a definition of course is that it is very hard to tell which f are in $\mathcal{D}$. So the following definition is important.

DEFINITION 2.23. A subspace $\mathcal{D}_0\subset\mathcal{D}$ is a *core* for L if L is the closure of its restriction to $\mathcal{D}_0$.

In particular, one can check invariance of a measure on the core:

LEMMA 2.24 (Prop. I.2.13 of [**86**]). *If $\mathcal{D}_0$ is a core for L, then a probability measure μ is invariant for the semigroup P_t, i.e. $P_t^*\mu=\mu$ if and only if $\int Lf\,d\mu = 0$ for all $f\in\mathcal{D}_0$.*

A function on the state space $\{0,1\}^{\mathbb{Z}}$ is called *local* if it only depends on finitely many coordinates. For ASEP we have the important

LEMMA 2.25 (Thm. I.3.9 of [**86**]). *The subspace $\mathcal{D}_0$ of local functions is a core for the generator (206).*

Hence we can check the following

PROPOSITION 2.26. *For any $\rho \in [0,1]$ the Bernoulli product measure π_ρ on $\{0,1\}^{\mathbb{Z}}$ with $\pi_\rho(\eta(x)=1)=\rho$ and $\pi_\rho(\eta(x)=0)=1-\rho$ is invariant for ASEP.*

PROOF. Let f be a local function, depending on $\eta(x)$, $|x| \leq B$. We want to show $\int Lf d\pi_\rho = 0$. We can assume $\int f d\pi_\rho = 0$, for otherwise we can subtract a constant to make it so, without affecting Lf. Changing η to $\eta^{x,x+1}$ we have

$$\begin{aligned}&\int \eta(x)(1-\eta(x+1))f(\eta^{x,x+1})d\pi_\rho(\eta)\\ &= \int \eta(x+1)(1-\eta(x))f(\eta)\frac{d\pi_\rho(\eta^{x,x+1})}{d\pi_\rho(\eta)}d\pi_\rho(\eta)\\ &= \int \eta(x+1)(1-\eta(x))f(\eta)d\pi_\rho(\eta)\end{aligned}$$

since $\frac{d\pi_\rho(\eta^{x,x+1})}{d\pi_\rho(\eta)} = 1$. Hence

(208)

$$\int Lf d\pi_\rho = \int (p-q)\sum_x (\eta(x+1)(1-\eta(x)) - \eta(x)(1-\eta(x+1)))f(\eta)d\pi(\eta).$$

Note that the summation is telescoping. Hence it can be rewritten as $\int gf d\pi_\rho$ where g does not depend on the variables $\eta(x)$, $|x| \leq B$. Since π_ρ is a product measure $\int gf d\pi_\rho = \int g d\pi_\rho \int f d\pi_\rho$ which vanishes by assumption. □

The π_ρ are the extremals of the set of translation invariant probability measures invariant for ASEP. There are other invariant measures which are not translation invariant, e.g. the *blocking measures* which are product measures with $\mu(\eta(x)=1) = \frac{(p/q)^x}{1+(p/q)^x}$.

2.14.1. *Height function.* Define $\hat\eta(x) = 2\eta(x) - 1$ which take values $\{-1,1\}$ instead of $\{0,1\}$ and define the height function of ASEP by

$$h(t,x) = \begin{cases} 2N(t) + \sum_{0<y\leq x}\hat\eta(t,y), & x>0,\\ 2N(t), & x=0,\\ 2N(t) - \sum_{x<y\leq 0}\hat\eta(t,y), & x<0,\end{cases} \tag{209}$$

where $N(t)$ is the net number of particles which crossed from site 1 to 0 up to time t, meaning that particle jumps $1 \to 0$ are counted as $+1$ and jumps $0 \to 1$ are counted as -1. We usually think of $h(x)$ in terms of its linear interpolation on $\mathbb{R}$. The reason for the funny definition with the $N(t)$ is that, defined this way, the $h(t,x)$ is the Markov process with state space simple random walk paths with the very simple dynamics that local minima jump to local maxima at rate q and local maxima jump to local minima at rate p.

2.15. Weakly asymmetric limit of simple exclusion. We will consider the weakly asymmetric simple exclusion with

$$q - p = \epsilon^{1/2} \tag{210}$$

We have to choose some parameters carefully to make the proof work. Then we will explain them. Let

$$\nu_\epsilon = p + q - 2\sqrt{qp} = \tfrac{1}{2}\epsilon + \tfrac{1}{8}\epsilon^2 + \mathcal{O}(\epsilon^3) \tag{211}$$

and

$$\lambda_\epsilon = \tfrac{1}{2}\log(q/p) = \epsilon^{1/2} + \tfrac{1}{3}\epsilon^{3/2} + \mathcal{O}(\epsilon^{5/2}). \tag{212}$$

Recall the definition (209) of the height function. The rescaled height function is

$$h_\epsilon(t,x) := \lambda_\epsilon h(\epsilon^{-2}t, \epsilon^{-1}x) + \epsilon^{-2}\nu_\epsilon t \tag{213}$$

The main result is that $h_\epsilon(t,x)$ converge to the Hopf-Cole solution of KPZ. What this means of course is that

$$z_\epsilon(t,x) = e^{-h_\epsilon(t,x)} \tag{214}$$

converges to the solution of the stochastic heat equation. This was first proved by Bertini and Giacomin [**15**] in the case of data close to equilibrium, and extended to the case of step initial data (Dirac initial data for the stochastic heat equation) in [**3**]. For step initial data one has to multiply $z_\epsilon(t,x)$ by a large factor $\epsilon^{-1/2}$ in order to get the initial Dirac in the limit. Since the stochastic heat equation is linear, one has a fair amount of freedom to do such things.

We now state the main results. $z_\epsilon(t,x)$ really lives on $\mathbb{R}_+ \times \epsilon\mathbb{Z}$, but we can extend it easily to $\mathbb{R}_+ \times \mathbb{R}$ by putting in little diagonal lines. We still call it $z_\epsilon(t,x)$. It is not continuous in t, but the jumps are small. Usually, convergence of discontinuous stochastic processes is done on the Skorohod space D of right continuous paths with left limits. But our jumps are small, and vanish in the limit, so the Skorohod topology is unnecessary. We can just use the uniform topology on D. We call the resulting space D_u. So our process lives in $D_u([0,\infty); C(\mathbb{R}))$. We have such a process $z_\epsilon(t,x)$ for each $\epsilon > 0$ and its distribution is a probability measure P_ϵ on $D_u([0,\infty); C(\mathbb{R}))$. The initial distibutions μ_ϵ, $\epsilon \in (0, 1/4)$ are the distributions of $h_\epsilon(0,x)$ or $z_\epsilon(0,x)$ under P_ϵ.

THEOREM 2.27 ([**15**]). *Suppose that the initial distribution μ_ϵ satisfies*

(1) *μ_ϵ converge weakly to a limit μ supported on continuous functions.*

(2) *For each* $n = 1, 2, \ldots$ *there is an* $a = a_n$ *and* $c = c_n$ *such that for all* $x \in \mathbb{R}$.

$$E_{\mu_\epsilon}[z^n(0,x)] \le ce^{a|x|} \tag{215}$$

and

$$E_{\mu_\epsilon}[|h(0,x) - h(0,y)|^{2n}] \le ce^{a(|x|+|y|)}(|x-y|). \tag{216}$$

Then P_ϵ, *for* $\epsilon \in (0, 1/4)$, *are a tight family of measures and the unique limit point is supported on* $C((0,\infty); C(\mathbb{R}))$ *(continuous in both space and time) and corresponds to the solution of the SHE with initial data* $z(0,x)$ *distributed as* μ.

Taking initial data Bernoulli 1/2 we have a stationary process $u_\epsilon(t,x) = \epsilon^{-1}(h_\epsilon(t, x+\epsilon) - h_\epsilon(t,x))$ for each $\epsilon \in (0, 1/4)$. The convergence of the h_ϵ is interpreted as distributional convergence of the u_ϵ. The limiting process $u(t,x)$ means nothing other than the distributional derivative in x of the limiting process $h(t,x)$. Since it is the limit of stationary processes, it is stationary itself. In this way, [**15**] prove that white noise is stationary for the stochastic Burgers equation. Remarkably, this is the only proof known.

The case of step initial data is not included in Theorem 2.27. Here μ_ϵ is concentrated on the special configuration with all zero's to the left of the origin and all one's to the right. The initial height function is $|x|$. So $h_\epsilon(0,x) = \lambda_\epsilon|\epsilon^{-1}x| \sim \epsilon^{-1/2}|x|$. Hence $z_\epsilon(t,x)$ has to be redefined as

$$z_\epsilon(t,x) = \epsilon^{-1/2}e^{-h_\epsilon(t,x)}. \tag{217}$$

We then have for the corresponding P_ϵ,

THEOREM 2.28 ([**3**]). P_ϵ, $\epsilon \in (0, 1/4)$, *are a tight family of measures and the unique limit point* P *is supported on* $C((0,\infty); C(\mathbb{R}))$ *and corresponds to the solution of the SHE with initial data* $z(0,x) = \delta_0(x)$.

The proof in [**3**] reduces the problem to the setup in [**15**] after a short time. Unfortunately, the proof of the key bound (2.29) in [**15**] is very hard to follow. We will concentrate here on providing a simple straightforward proof, and to do this we will work on a circle geometry $\epsilon\mathbb{Z}/\mathbb{Z}$. We always choose $\epsilon^{-1} \in \mathbb{Z}$ and we fix the number of particles to be $\epsilon^{-1}/2$ so that the height function is properly periodic.

2.15.1. *Gärtner's microscopic Hopf-Cole transformation.* We write the stochastic differential equation which governs the evolution of $z_\epsilon(t,x)$.

$$dz_\epsilon = \Omega_\epsilon z_\epsilon dt + (e^{-2\lambda_\epsilon} - 1)z_\epsilon dM_\epsilon^- + (e^{2\lambda_\epsilon} - 1)z_\epsilon dM_\epsilon^- \tag{218}$$

where

$$\Omega = \epsilon^{-2}\nu + (e^{-2\lambda} - 1)r_- + (e^{2\lambda} - 1)r_+ \tag{219}$$

and $dM_\pm(t,x) = dP_\pm(t,x) - r_\pm(x)dt$ where $P_-(t,x), P_+(t,x)$, $x \in \epsilon\mathbb{Z}$ are independent Poisson processes running at rates $r_-(t,x), r_+(t,x)$.

$$r_-(x) = \epsilon^{-2}q(1-\eta(x))\eta(x+1) = \frac{1}{4}\epsilon^{-2}q(1-\hat{\eta}(x))(1+\hat{\eta}(x+1)) \tag{220}$$

$$r_+(x) = \epsilon^{-2}p\eta(x)(1-\eta(x+1)) = \frac{1}{4}\epsilon^{-2}p(1+\hat{\eta}(x))(1-\hat{\eta}(x+1)) \tag{221}$$

This is just a way of writing that the process jumps at rates

$$r_-(x) \tag{222}$$

to $e^{-2\lambda_\epsilon}z_\epsilon$ and

$$r_+(x) \tag{223}$$

to $e^{2\lambda_\epsilon}z_\epsilon$, independently at each site $x \in \epsilon\mathbb{Z}$.

Let

$$D_\epsilon = 2\sqrt{pq} = 1 - \frac{1}{2}\epsilon + \mathcal{O}(\epsilon^2) \tag{224}$$

and $\Delta_\epsilon f(x) = \epsilon^{\ 2}(f(x+\epsilon) - 2f(x) + f(x-\epsilon))$ be the $\epsilon\mathbb{Z}$ Laplacian. We also have

$$\tfrac{1}{2}D_\epsilon\Delta_\epsilon z_\epsilon = \tfrac{1}{2}\epsilon^{-2}D_\epsilon(e^{-\lambda_\epsilon\hat{\eta}(x+1)} - 2 + e^{\lambda_\epsilon\hat{\eta}(x)})z_\epsilon \tag{225}$$

The key point is that parameters can be chosen so that

$$\Omega_\epsilon = \tfrac{1}{2}\epsilon^{-2}D_\epsilon(e^{-\lambda_\epsilon\hat{\eta}(x+1)} - 2 + e^{\lambda_\epsilon\hat{\eta}(x)}) \tag{226}$$

We can do this because the four cases, corresponding to the four possibilities for $\hat{\eta}(x), \hat{\eta}(x+1)$: 11, $(-1)(-1)$, $1(-1)$, $(-1)1$, give four equations in three unknowns,

$$\begin{array}{rl}
11 & \tfrac{1}{2}\epsilon^{-2}D_\epsilon(e^{-\lambda_\epsilon} - 2 + e^{\lambda_\epsilon}) = \epsilon^{-2}\nu_\epsilon \\
(-1)(-1) & \tfrac{1}{2}\epsilon^{-2}D_\epsilon(e^{\lambda_\epsilon} - 2 + e^{-\lambda_\epsilon}) = \epsilon^{-2}\nu_\epsilon \\
1(-1) & \tfrac{1}{2}\epsilon^{-2}D_\epsilon(e^{\lambda_\epsilon} - 2 + e^{\lambda_\epsilon}) = \epsilon^{-2}\nu_\epsilon + (e^{2\lambda_\epsilon} - 1)\epsilon^{-2}p \\
(-1)1 & \tfrac{1}{2}\epsilon^{-2}D_\epsilon(e^{-\lambda_\epsilon} - 2 + e^{-\lambda_\epsilon}) = \epsilon^{-2}\nu_\epsilon + (e^{-2\lambda_\epsilon} - 1)\epsilon^{-2}q
\end{array}$$

Luckily, the first two equations are the same, so it is actually three equations in three unknowns, with solution given by (212), (211) and (224).

Hence [**65**],[**15**],

$$dz_\epsilon = \tfrac{1}{2}D_\epsilon\Delta_\epsilon z_\epsilon + z_\epsilon dM_\epsilon \tag{227}$$

where

$$dM_\epsilon(x) = (e^{-2\lambda_\epsilon} - 1)dM_-(x) + (e^{2\lambda_\epsilon} - 1)dM_+(x) \tag{228}$$

are martingales in t with

$$d\langle M_\epsilon(x), M_\epsilon(y)\rangle = \epsilon^{-1}\mathbf{1}(x=y)b_\epsilon(\tau_{-[\epsilon^{-1}x]}\eta)dt \tag{229}$$

where $\tau_x\eta(y) = \eta(y-x)$ and

$$b_\epsilon(\eta) = 1 - \hat{\eta}(1)\hat{\eta}(0) + \hat{b}_\epsilon(\eta) \tag{230}$$

where

$$\begin{aligned}\hat{b}_\epsilon(\eta) = \epsilon^{-1}\{&[p((e^{-2\lambda_\epsilon}-1)^2-4\epsilon)+q((e^{2\lambda_\epsilon}-1)^2-4\epsilon)]\\ &+[q(e^{-2\lambda_\epsilon}-1)^2-p(e^{2\lambda_\epsilon}-1)^2](\hat{\eta}(1)-\hat{\eta}(0))\\ &-[q(e^{-2\lambda_\epsilon}-1)^2+p(e^{2\lambda_\epsilon}-1)^2-\epsilon]\hat{\eta}(1)\hat{\eta}(0)\}.\end{aligned} \tag{231}$$

Clearly $b_\epsilon \geq 0$. It is easy to check that there is a $C < \infty$ such that

$$0 \leq \hat{b}_\epsilon \leq C\epsilon^{1/2} \tag{232}$$

and, for sufficiently small $\epsilon > 0$,

$$0 \leq b_\epsilon \leq 3. \tag{233}$$

We have the following bound on the initial data. For each $p = 1, 2, \ldots$ there exists $C_p < \infty$ such that for all $x \in \mathbb{R}$,

$$E[z_\epsilon^p(0,x)] \leq e^{C_p|x|}. \tag{234}$$

For any $\delta > 0$ let $\mathscr{P}_\epsilon^\delta$ denote the distribution of $z_\epsilon(t,x)$, $t \in [\delta,\infty)$, as measure on $D[[\delta,\infty), C(\mathbb{R})]$ where D means the right continuous paths with left limits, with the topology of uniform convergence on compact sets. In [**15**], Section 4 it is shown that if (234) holds, then, for any $\delta > 0$, $\mathscr{P}_\epsilon^\delta$, $\epsilon > 0$ are tight. The limit points are consistent as $\delta \searrow 0$, and from the integral version of (227),

$$\begin{aligned}z_\epsilon(t,x) = {}&\epsilon\sum_{y\in\epsilon\mathbb{Z}} p_\epsilon(t,x-y)z_\epsilon(0,y)\\ &+\int_0^t \epsilon\sum_{y\in\epsilon\mathbb{Z}} p_\epsilon(t-s,x-y)z_\epsilon(s,y)dM_\epsilon(s,y)\end{aligned} \tag{235}$$

where $p_\epsilon(t,x)$ are the transition probabilities for the continuous time random walk with generator $\frac{1}{2}D\Delta_\epsilon$, normalized so that $p_\epsilon(t,x) \to p(t,x) = \frac{e^{-x^2/2t}}{\sqrt{2\pi t}}$, we have

$$\lim_{t\to 0}\lim_{\epsilon\to 0} E\left[\left(z_\epsilon(t,x) - \epsilon\sum_{y\in\epsilon\mathbb{Z}} p_\epsilon(t,x-y)z_\epsilon(0,y)\right)^2\right] = 0 \tag{236}$$

so that the initial data hold under the limit $\mathscr{P}$. Finally, we need to identify the limit of the martingale terms. Let φ be a smooth test function on $\mathbb{R}$. We hope to show that under $\mathscr{P}$, (154) and (155) are local martingales. We have that

$$N_{t,\epsilon}(\varphi) := \epsilon\sum_{x\in\epsilon\mathbb{Z}} \varphi(x)z_\epsilon(t,x) - \frac{1}{2}\int_0^t \epsilon\sum_{x\in\epsilon\mathbb{Z}} D_\epsilon\Delta_\epsilon\varphi(x)z_\epsilon(s,x)ds \tag{237}$$

is a martingale under P_ϵ. So the key point is to identify the quadratic process. We have that

$$\Lambda_{t,\epsilon}(\varphi) := N_{t,\epsilon}(\varphi)^2 - \tfrac{1}{2}\int_0^t \epsilon \sum_{\epsilon\mathbb{Z}} \varphi^2(x) b_\epsilon(s,x) z_\epsilon^2(s,x)ds \tag{238}$$

is a martingale under P_ϵ.

The problem with (238) is that we have the b_ϵ instead of 1. First let's note that even if we don't know what the asymptotic behaviour of b_ϵ is, we do know $|b_\epsilon| \leq 3$. Using this, we can recover all the estimates we had before about modulus of continuity. In particular, we have the tightness. But we still need to identify the limit martingales. Let

$$a_\epsilon(t,x) = \hat{\eta}(\epsilon^{-2}t, x+1)\hat{\eta}(\epsilon^{-2}t, x) z_\epsilon^2(t,x) \tag{239}$$

and recall that $b_\epsilon z_\epsilon^2 = z_\epsilon^2 + a_\epsilon$.

So the main estimate needed to identify the limit with martingales is

Proposition 2.29 (Bertini-Giacomin [**15**]). *Let*

$$a_\epsilon(t,x) = \hat{\eta}(\epsilon^{-2}t, x+1)\hat{\eta}(\epsilon^{-2}t, x) z_\epsilon^2(t,x) \tag{240}$$

Then for any smooth test function $\varphi \in C_0^\infty(\mathbb{R})$ under the weakly asymmetric exclusion P_ϵ,

$$\int_0^t \epsilon \sum_{x\in\epsilon\mathbb{Z}} \varphi^2(x) a_\epsilon(s,x)ds \to 0 \tag{241}$$

Note that $a_\epsilon(t,x)$ is of order 1. The proposition is a kind of ergodic theorem. We expect essentially product measure with density 1/2 under which $\hat{\eta}(x+1)$ and $\hat{\eta}(x)$ are independent and mean zero. The term $z_\epsilon^2(t,x)$ shouldn't mess things up too much because we could replace it by $z_\epsilon^2(t, x-\epsilon)$. The error would be of order $\epsilon^{1/2}$, so it doesn't matter. Then, in equilibrium, $z_\epsilon^2(t, x-\epsilon)$ is independent of $\hat{\eta}(x+1)$ and $\hat{\eta}(x)$.

Proof. We will give a different proof from the one of Bertini and Giacomin, which comes from hydrodynamic limits. Let μ_ϵ be the probability measure on $\{0,1\}^{\epsilon\mathbb{Z}/\mathbb{Z}}$ corresponding to our random initial data. Let ν be the uniform measure on configurations with $\epsilon^{-1}/2$ particles. Because we are on $\epsilon\mathbb{Z}/\mathbb{Z}$ there are ϵ^{-1} sites and

$$H(\mu_\epsilon/\nu) := E_{\mu_\epsilon}[\log(\mu_\epsilon/\nu)] \leq C\epsilon^{-1}. \tag{242}$$

Let E_ϵ denote the expectation with respect to the process starting from μ_ϵ and E_ϵ^{eq} denote the expectation with respect to the equilibrium process starting with π. We have the entropy inequality,

$$E_\epsilon[f_\epsilon] \leq K^{-1}\epsilon \log E_\epsilon^{\text{eq}}[e^{K\epsilon^{-1}f_\epsilon}] + K^{-1}\epsilon H(\mu_\epsilon/\nu). \tag{243}$$

Hence if for any $K \in \mathbb{R}$,

$$\limsup_{\epsilon \to 0} \epsilon \log E_\epsilon^{\mathrm{eq}}[e^{K\epsilon^{-1} f_\epsilon}] = 0 \tag{244}$$

we have $E_\epsilon[f_\epsilon] \to 0$.

Our f will be of the form $\epsilon \int_0^s V(\eta(s))ds$. So we have to estimate $\epsilon \log \int u(t,\eta) d\pi_{1/2}(\eta)$ where

$$u(t,\eta) = E_\epsilon \left[\exp\left\{ \int_0^t V(\eta(s))ds \right\} \mid \eta(0) = \eta \right]. \tag{245}$$

By Jensen's inequality,

$$\log \int u d\nu \le \tfrac{1}{2} \log \int u^2 d\nu. \tag{246}$$

By the Feynman-Kac formula u solves

$$\partial_t u = Lu + Vu. \tag{247}$$

Multiplying by u and integrating by parts we get

$$\tfrac{1}{2t} \log \int u^2 d\nu = \frac{\frac{1}{t}\int_0^t \left\{ \int V u^2 d\nu - \mathscr{D}(u) \right\} ds}{\int u^2 d\nu} \tag{248}$$

where $\mathscr{D}(u)$ is the Dirichlet form $\epsilon^{-2}\frac{1}{2}\sum_x \int |\nabla_{x,x+1} u|^2 d\nu$ with $\nabla_{x,x+1} u(\eta) = u(\eta^{x,x+1}) - u(\eta)$, $\eta^{x,x+1}$ being the configuration with the occupation variables exchanged between sites x and $x+1$. By the Rayleigh-Ritz formula the last ratio is bounded above by the principle eigenvalue of $S+V$ where $S = L + L^*$ is the generator of the symmetric simple exclusion process (i.e. $p = q$). Thus the problem (244) is reduced to large deviations for a reversible process.

The following estimate combines the one and two block estimates of hydrodynamic limits. Let f be a local function, and let $\bar{f}(\rho) = E_{\pi_\rho}[f]$, where π_ρ is a the product measure with density ρ. Let

$$V_{\epsilon,\delta} = \sum_{x \in \epsilon\mathbb{Z}} \varphi^2(x) \left[f(\tau_x \eta) - \bar{f}(\bar{\eta}_{x,\epsilon^{-1}\delta}) \right] \tag{249}$$

where $\bar{\eta}_{x,\epsilon^{-1}\delta}$ is the empirical density in an interval of length $e^{-1}\delta$ around x. Then

$$\lim_{\delta \to 0} \lim_{\epsilon \to 0} \sup_{\|u\|_2 = 1} \left\{ \int V_{\epsilon,\delta} u^2 d\nu - \mathscr{D}_\epsilon(u) \right\} = 0 \tag{250}$$

We will be using this with

$$f(\eta) = \hat{\eta}(1)\hat{\eta}(0) \tag{251}$$

for which it is easy to see that $\bar{f}(\rho) = (\rho - \frac{1}{2})^2$. Now note that we have apriori that $z(t,x)$ is Hölder $\frac{1}{2}-$ in space. So we can have the same estimate (250) if we replace (249) by

$$V_{\epsilon,\delta} = \sum_{x\in\epsilon\mathbb{Z}} \varphi^2(x) \left[f(\tau_x\eta) - \bar{f}(\bar{\eta}_{x,\epsilon^{-1}\delta})\right] z_\delta^2(t,x). \tag{252}$$

where

$$z_\delta^2(t,x) = \min\left\{z^2(t, x + \tfrac{1}{2}\epsilon^{-1}\delta), z^2(t, x + \tfrac{1}{2}\epsilon^{-1}\delta)\right\}. \tag{253}$$

The big conclusion of all this hydrodynamics argument is that if we want to prove that

$$\limsup_{\epsilon\to 0} P_\epsilon\Big(\int_0^t \epsilon \sum_{x\in\epsilon\mathbb{Z}} \varphi^2(x) a_\epsilon(s,x) ds \geq \lambda\Big) = 0 \tag{254}$$

it suffices to prove that

$$\limsup_{\delta\to 0}\limsup_{\epsilon\to 0} P_\epsilon\Big(\int_0^t \epsilon \sum_{x\in\epsilon\mathbb{Z}} \varphi^2(x) (\bar{\eta}_{x,\epsilon^{-1}\delta})^2 z_\delta^2(t,x) ds \geq \lambda\Big) = 0. \tag{255}$$

To prove this, note that

$$\bar{\eta}_{x,\epsilon^{-1}\delta} = \epsilon^{1/2}\delta^{-1}(\log z_\epsilon(x + \tfrac{1}{2}\epsilon^{-1}\delta) - \log z_\epsilon(x - \tfrac{1}{2}\epsilon^{-1}\delta)) \tag{256}$$

so that

$$\epsilon \sum_{x\in\epsilon\mathbb{Z}} \varphi^2(x)(\bar{\eta}_{x,\epsilon^{-1}\delta})^2 z_\delta^2(t,x) \leq 2\delta^{-2}\epsilon^2 \sum_{x\in\epsilon\mathbb{Z}} \varphi_\delta^2(x)(\log z_\epsilon(x))^2 z_\epsilon^2(x) \tag{257}$$

where $\varphi_\delta(x) = (\varphi_\delta^2(x + \frac{1}{2}\epsilon^{-1}\delta) + \varphi_\delta^2(x - \frac{1}{2}\epsilon^{-1}\delta))^{1/2}$. The last term clearly goes to zero under P_ϵ as $\epsilon \to 0$, for any fixed δ (there is an extra ϵ in front.) This completes the proof.

□

2.15.2. *Invariance of Brownian motion for KPZ.* As pointed out in [**15**], since the Bernoulli product measure is invariant for ASEP, taking the weakly asymmetric limit in equilibrium one obtains a stationary process for the distributional derivative $u = \partial_x h$ of the KPZ solution. In particular, white noise is invariant for the Hopf-Cole solution of the stochastic Burgers equation. Note that u determines h up to a constant. We can think of equivalence classes of height functions h where two height functions are equivalent if they differ by an absolute constant. In [**63**] it is observed that in (41) all terms except the first, $\partial_t h_\kappa$, only depend on the equivalence class of h_κ, and therefore through this equation, the absolute height can be recovered. In this way, it is shown that a two sided Brownian motion with a height shift given by (an independent) Lebesgue measure is invariant for the KPZ equation.

2.16. Steepest descent analysis of the Tracy-Widom step Bernoulli formula. We only give here a heuristic explanation of the proof. We are taking the limit of (96) with $\rho = 1/2$, $t = \epsilon^{-3/2}T$, $x = \epsilon^{-1}X$, $q - p = \gamma = \epsilon^{1/2}$, and

$$m = \tfrac{1}{2}\left[-s\epsilon^{-1/2} + \tfrac{t}{2} + x\right]. \tag{258}$$

The first term in the integrand of (96) is the infinite product $\prod_{k=0}^{\infty}(1-\mu\tau^k)$. Observe that $\tau = q/p \approx 1 - 2\epsilon^{1/2}$ and that $S_{\tau+}$, the contour on which μ lies, is a circle centered at zero of radius between τ and 1. The infinite product is not well behaved along most of this contour, however we can deform the contour to one along which the product is not highly oscillatory. However, the Fredholm determinant has poles at every $\mu = \tau^k$ and the deformation must avoid passing through them. Observe that if

$$\mu = \epsilon^{1/2}\tilde{\mu} \tag{259}$$

then

$$\prod_{k=0}^{\infty}(1-\mu\tau^k) \approx e^{-\sum_{k=0}^{\infty}\mu\tau^k} = e^{-\mu/(1-\tau)} \approx e^{-\tilde{\mu}/2}.$$

We make the $\mu \mapsto \epsilon^{-1/2}\tilde{\mu}$ change of variables and find that if we consider a $\tilde{\mu}$ contour

$$\tilde{C}_\epsilon = \{e^{i\theta}\}_{\pi/2\le\theta\le 3\pi/2} \cup \{x \pm i\}_{0<x\le \epsilon^{-1/2}-1} \cup \{\epsilon^{-1/2} - 1 + iy\}_{-1<y<1},$$

then the above approximations are reasonable. Thus the infinite product goes to $\exp\{-\tilde{\mu}/2\}$.

Now we turn to the Fredholm determinant and determine a candidate for the pointwise limit of the kernel. The kernel $J_\mu(\eta,\eta')$ is given by an integral whose integrand has four main components: An exponential

$$\exp\{\Lambda(\zeta) - \Lambda(\eta')\},$$

a rational function (we include the differential with this term for scaling purposes)

$$d\zeta/\eta'(\zeta-\eta),$$

a doubly infinite sum

$$\mu f(\mu, \zeta/\eta'),$$

and an infinite product

$$g(\eta')/g(\zeta).$$

We proceed by the method of steepest descent, so in order to determine the region along the ζ and η contours which affects the asymptotics we must consider the exponential term first. The argument of the exponential is given by $\Lambda(\zeta) - \Lambda(\eta')$ where

$$\Lambda(\zeta) = -x\log(1-\zeta) + \frac{t\zeta}{1-\zeta} + m\log(\zeta),$$

where x, t and m are as in (258). For small ϵ, $\Lambda(\zeta)$ has a critical point in an $\epsilon^{1/2}$ neighborhood of -1. One has some freedom in where to center the expansion and for purposes of having a nice ultimate answer we choose to use

$$\xi = -1 - 2\epsilon^{1/2}\frac{X}{T}$$

We can rewrite the argument of the exponential as $(\Lambda(\zeta) - \Lambda(\xi)) - (\Lambda(\eta') - \Lambda(\xi)) = \Psi(\zeta) - \Psi(\eta')$. The idea of extracting asymptotics for this term (which starts like those done in [**116**] but quickly becomes more involved due to the fact that τ tends to 1 as ϵ goes to zero) is then to deform the ζ and η contours to lie along curves such that outside the scale $\epsilon^{1/2}$ around ξ, $\mathrm{Re}\Psi(\zeta)$ is very negative, and $\mathrm{Re}\Psi(\eta')$ is very positive, and hence the contribution from those parts of the contours is negligible. Rescaling around ξ to blow up this $\epsilon^{1/2}$ scale, gives us the asymptotic exponential term. This change of variables sets the scale at which we should analyze the other three terms in the integrand for the J kernel.

Returning to $\Psi(\zeta)$, we make a Taylor expansion around ξ and find that in a neighborhood of ξ

$$\Psi(\zeta) \approx -\frac{T}{48}\epsilon^{-3/2}(\zeta - \xi)^3 + \tfrac{1}{2}(s + \tfrac{X^2}{2T})\epsilon^{-1/2}(\zeta - \xi).$$

This suggests the following change of variables

(260)

$$\tilde{\zeta} = 2^{-4/3}\epsilon^{-1/2}(\zeta - \xi) \qquad \tilde{\eta} = 2^{-4/3}\epsilon^{-1/2}(\eta - \xi) \qquad \tilde{\eta}' = 2^{-4/3}\epsilon^{-1/2}(\eta' - \xi),$$

after which our Taylor expansion takes the form

$$\Psi(\tilde{\zeta}) \approx -\frac{T}{3}\tilde{\zeta}^3 + 2^{1/3}(s + \tfrac{X^2}{2T})\tilde{\zeta}.$$

In the spirit of steepest descent analysis we would like the ζ contour to leave ξ in a direction where this Taylor expansion is decreasing rapidly. This is accomplished by leaving at an angle $\pm 2\pi/3$. Likewise, since $\Psi(\eta)$ should increase rapidly, η should leave ξ at angle $\pm\pi/3$. Since $\rho_+ = 1/2$, $\alpha = 1$ which means that the ζ contour is originally on a circle of diameter $[-1+\delta, 1+\delta]$ and the η contour on a circle of diameter $[-1+2\delta, 1-\delta]$ for some positive δ (which can and should depend on ϵ so as to ensure that $|\zeta/\eta| \in (1, \tau^{-1})$). In order to deform these contours to their steepest descent contours without changing the value of the determinant, great care must be taken to avoid the poles of f, which occur whenever $\zeta/\eta' = \tau^k$, $k \in \mathbb{Z}$, and the poles of $1/g$, which occur whenever $\zeta = -\tau^{-n}$, $n \geq 0$. We will ignore these considerations in the formal calculation. The one very important consideration in this deformation, even formally, is that we must end up with contours which lie to the right of the poles of the $1/g$ function.

Let us now assume that we can deform our contours to curves along which Ψ rapidly decays in ζ and increases in η, as we move along them away from ξ. If we apply the change of variables in (260) the straight part of our contours become infinite rays at angles $\pm 2\pi/3$ and $\pm\pi/3$ which we call $\tilde{\Gamma}_\zeta$ and $\tilde{\Gamma}_\eta$[9].

Applying this change of variables to the kernel of the Fredholm determinant changes the L^2 space and hence we must multiply the kernel by the Jacobian term $2^{4/3}\epsilon^{1/2}$. We will include this term with the $\mu f(\mu, z)$ term and take the $\epsilon \to 0$ limit of that product.

Before we consider that term, however, it is worth looking at the new infinite product term $g(\eta')/g(\zeta')$. In order to do that let us consider the following. Set

$$q = 1 - r, \qquad a = \frac{\log \alpha(c - xr)}{\log q}, \qquad b = \frac{\log \alpha(c - yr)}{\log q}.$$

Then observe that

$$\prod_{n=0}^{\infty} \frac{1 + (1-r)^n \alpha(-c + xr)}{1 + (1-r)^n \alpha(-c + yr)} = \frac{(q^a; q)_\infty}{(q^b; q)_\infty} = \frac{\Gamma_q(b)}{\Gamma_q(a)}(1-q)^{b-a}$$
$$= \frac{\Gamma_q(b)}{\Gamma_q(a)} e^{(b-a)\log r} = \frac{\Gamma_{1-r}(-r^{-1}\log(\alpha c) + c^{-1}y + o(r))}{\Gamma_{1-r}(-r^{-1}\log(\alpha c) + c^{-1}x + o(r))}$$
$$e^{c^{-1}(y-x)\log r + o(r \log r)},$$

(261)

where the q-Gamma function and the q-Pochhammer symbols are given by

$$\Gamma_q(x) := \frac{(q;q)_\infty}{(q^x;q)_\infty}(1-q)^{1-x}$$

when $|q| < 1$ and

$$(a;q)_\infty = (1-a)(1-aq)(1-aq^2)\cdots.$$

The notation $o(f(r))$ above refers to a function $f'(r)$ such that $f'(r)/f(r) \to 0$ as $r \to 0$. The q-Gamma function converges to the usual Gamma function as $q \to 1$, uniformly on compact sets (see [**5**]).

Now consider the g terms and observe that in the rescaled variables this corresponds with (261) with $r = 2\epsilon^{1/2}$, $c = 1$ (recall $\alpha = 1$ as well) and

$$y = 2^{1/3}\tilde{\zeta} - \frac{X}{T}, \qquad x = 2^{1/3}\tilde{\eta}' - \frac{X}{T}$$

Since $\alpha c = 1$ and since we are away from the poles and zeros of the Gamma functions we find that

$$\frac{g(\eta')}{g(\zeta)} \to \frac{\Gamma\left(2^{1/3}\tilde{\zeta} - \frac{X}{T}\right)}{\Gamma\left(2^{1/3}\tilde{\eta}' - \frac{X}{T}\right)} \exp\left\{2^{1/3}(\tilde{\zeta} - \eta')\log(2\epsilon^{1/2})\right\}. \tag{262}$$

[9]Note that this is *not* the actual definition of the contours which one uses in the real proof because of the singularity problem.

This exponential can be rewritten as

$$\exp\left\{\frac{\tilde{z}}{4}\log\epsilon\right\}\exp\left\{2^{1/3}\log(2)(\tilde{\zeta}-\tilde{\eta}')\right\}. \tag{263}$$

where

$$\tilde{z} = 2^{4/3}(\tilde{\zeta}-\tilde{\eta}'). \tag{264}$$

It appears that there is a problem in these asymptotics as ϵ goes to zero, however we will find that this apparent divergence exactly cancels with a similar term in the doubly infinite summation term asymptotics. We will now show how that $\log\epsilon$ in the exponent can be absorbed into the $2^{4/3}\epsilon^{1/2}\mu f(\mu,\zeta/\eta')$ term. Recall

$$\mu f(\mu,z) = \sum_{k=-\infty}^{\infty}\frac{\mu\tau^k}{1-\tau^k\mu}z^k.$$

If we let $n_0 = \lfloor\log(\epsilon^{-1/2})/\log(\tau)\rfloor$ then observe that

$$\mu f(\mu,z) = \sum_{k=-\infty}^{\infty}\frac{\mu\tau^{k+n_0}}{1-\tau^{k+n_0}\mu}z^{k+n_0} = z^{n_0}\tau^{n_0}\mu\sum_{k=-\infty}^{\infty}\frac{\tau^k}{1-\tau^k\tau^{n_0}\mu}z^k.$$

By the choice of n_0, $\tau^{n_0}\approx\epsilon^{-1/2}$ so

$$\mu f(\mu,z)\approx z^{n_0}\tilde{\mu}f(\tilde{\mu},z).$$

The discussion on the exponential term indicates that it suffices to understand the behavior of this function only in the region where ζ and η' are within a neighborhood of ξ of order $\epsilon^{1/2}$. Equivalently, letting $z=\zeta/\eta'$, it suffices to understand $\mu f(\mu,z)\approx z^{n_0}\tilde{\mu}f(\tilde{\mu},z)$ for

$$z = \frac{\zeta}{\eta'} = \frac{\xi+2^{4/3}\epsilon^{1/2}\tilde{\zeta}}{\xi+2^{4/3}\epsilon^{1/2}\tilde{\eta}'}\approx 1-\epsilon^{1/2}\tilde{z}.$$

Let us now consider z^{n_0} using the fact that $\log\tau\approx-2\epsilon^{1/2}$:

$$z^{n_0}\approx(1-\epsilon^{1/2}\tilde{z})^{\epsilon^{-1/2}\frac{1}{4}\log\epsilon}\approx e^{-\frac{1}{4}\tilde{z}\log\epsilon}. \tag{265}$$

Plugging back in the value of $\tilde{z}$ in terms of $\tilde{\zeta}$ and $\tilde{\eta}'$ we see that this prefactor of z^{n_0} exactly cancels the $\log\epsilon$ term which came from the g infinite product term.

What remains is to determine the limit of $2^{4/3}\epsilon^{1/2}\tilde{\mu}f(\tilde{\mu},z)$ as ϵ goes to zero and for $z\approx 1-\epsilon^{1/2}\tilde{z}$. This limit can be found by interpreting the infinite sum as a Riemann sum approximation for an appropriate integral. Define $t=k\epsilon^{1/2}$, then observe that

$$\epsilon^{1/2}\tilde{\mu}f(\tilde{\mu},z) = \sum_{k=-\infty}^{\infty}\frac{\tilde{\mu}\tau^{t\epsilon^{-1/2}}z^{t\epsilon^{-1/2}}}{1-\tilde{\mu}\tau^{t\epsilon^{-1/2}}}\epsilon^{1/2}\to\int_{-\infty}^{\infty}\frac{\tilde{\mu}e^{-2t}e^{-\tilde{z}t}}{1-\tilde{\mu}e^{-2t}}dt.$$

This used the fact that $\tau^{t\epsilon^{-1/2}} \to e^{-2t}$ and that $z^{t\epsilon^{-1/2}} \to e^{-\tilde{z}t}$, which hold at least pointwise in t. If we change variables of t to $t/2$ and multiply the top and bottom by e^{-t} then we find that

$$2^{4/3}\epsilon^{1/2}\mu f(\mu,\zeta/\eta') \to 2^{1/3}\int_{-\infty}^{\infty}\frac{\tilde{\mu}e^{-\tilde{z}t/2}}{e^t-\tilde{\mu}}dt.$$

As far as the final term, the rational expression, under the change of variables and zooming in on ξ, the factor of $1/\eta'$ goes to -1 and the $\frac{d\zeta}{\zeta-\eta'}$ goes to $\frac{d\tilde{\zeta}}{\tilde{\zeta}-\tilde{\eta}'}$.

Therefore we formally find the following kernel: $-K_{a'}^{\mathrm{csc},\Gamma}(\tilde{\eta},\tilde{\eta}')$ acting on $L^2(\tilde{\Gamma}_\eta)$ where:

$$K_{a'}^{\mathrm{csc},\Gamma}(\tilde{\eta},\tilde{\eta}') = \int_{\tilde{\Gamma}_\zeta}\exp\{-\frac{T}{3}(\tilde{\zeta}^3-\tilde{\eta}'^3)+2^{1/3}a'(\tilde{\zeta}-\tilde{\eta}')\}2^{1/3}$$
$$\times\left(\int_{-\infty}^{\infty}\frac{\tilde{\mu}e^{-2^{1/3}t(\tilde{\zeta}-\tilde{\eta}')}}{e^t-\tilde{\mu}}dt\right)\frac{\Gamma\left(2^{1/3}\tilde{\zeta}-\frac{X}{T}\right)}{\Gamma\left(2^{1/3}\tilde{\eta}'-\frac{X}{T}\right)}\frac{d\tilde{\zeta}}{\tilde{\zeta}-\tilde{\eta}},$$

where $a' = s+\frac{X^2}{2T}+\log 2$ (recall that this $\log 2$ came from (263)).

We have the identity

$$\int_{-\infty}^{\infty}\frac{\tilde{\mu}e^{-\tilde{z}t/2}}{e^t-\tilde{\mu}}dt = (-\tilde{\mu})^{-\tilde{z}/2}\pi\csc(\pi\tilde{z}/2), \tag{266}$$

where the branch cut in $\tilde{\mu}$ is along the positive real axis, hence $(-\tilde{\mu})^{-\tilde{z}/2} = e^{-\log(-\tilde{\mu})\tilde{z}/2}$ where log is taken with the standard branch cut along the negative real axis. We may use the identity to rewrite the kernel as

$$K_{a'}^{\mathrm{csc},\Gamma}(\tilde{\eta},\tilde{\eta}') = \int_{\tilde{\Gamma}_\zeta}\exp\{-\frac{T}{3}(\tilde{\zeta}^3-\tilde{\eta}'^3)+2^{1/3}a'(\tilde{\zeta}-\tilde{\eta}')\}2^{1/3}$$
$$\times\frac{\pi(-\tilde{\mu})^{-2^{1/3}(\tilde{\zeta}-\tilde{\eta}')}}{\sin(\pi 2^{1/3}(\tilde{\zeta}-\tilde{\eta}'))}\frac{\Gamma\left(2^{1/3}\tilde{\zeta}-\frac{X}{T}\right)}{\Gamma\left(2^{1/3}\tilde{\eta}'-\frac{X}{T}\right)}\frac{d\tilde{\zeta}}{\tilde{\zeta}-\tilde{\eta}}.$$

To make this cleaner we replace $\tilde{\mu}/2$ with $\tilde{\mu}$. Taking into account this change of variables (it also changes the $\exp\{-\tilde{\mu}/2\}$ in front of the determinant to $\exp\{-\tilde{\mu}\}$), we find that a nice way to write the final answer is

$$\int_{\tilde{\mathcal{C}}} e^{-\tilde{\mu}}\frac{d\tilde{\mu}}{\tilde{\mu}}\det(I-K^{\mathrm{csc},\Gamma}_{s+\frac{X^2}{2T}})_{L^2(\tilde{\Gamma}_\eta)}.$$

The details can be found in [**42**].

The asymptotic analysis of the corner growth case $\rho = 1$ is slightly easier because there are fewer constraints on the contours, and no term (262). In the analysis though, there is apparently no term now to cancel the divergent term in (265). However, this is easily resolved by simply shifting the initial problem by a factor $\log(\epsilon^{-1/2})$. The nice thing is that this is exactly what we need (217) in order to see an initial Dirac delta function.

References

[1] Tom Alberts, Konstantin Khanin, and Jeremy Quastel. The continuum directed random polymer. 02 2012.

[2] Tom Alberts, Konstantin Khanin, and Jeremy Quastel. The intermediate disorder regime for directed polymers in dimension 1+1. 02 2012.

[3] Gideon Amir, Ivan Corwin, and Jeremy Quastel. Probability distribution of the free energy of the continuum directed random polymer in $1+1$ dimensions. *Comm. Pure Appl. Math.*, 64(4):466–537, 2011.

[4] Greg W. Anderson, Alice Guionnet, and Ofer Zeitouni. *An introduction to random matrices*, volume 118 of *Cambridge Studies in Advanced Mathematics*. Cambridge University Press, Cambridge, 2010.

[5] George E. Andrews, Richard Askey, and Ranjan Roy. *Special functions*, volume 71 of *Encyclopedia of Mathematics and its Applications*. Cambridge University Press, Cambridge, 1999.

[6] Antonio Auffinger and Michael Damron. A simplified proof of the relation between scaling exponents in first-passage percolation. *arXiv:1109.0523*, 2011.

[7] Jinho Baik, Percy Deift, and Kurt Johansson. On the distribution of the length of the longest increasing subsequence of random permutations. *J. Amer. Math. Soc.*, 12(4):1119–1178, 1999.

[8] Jinho Baik, Patrik L. Ferrari, and Sandrine Péché. Limit process of stationary TASEP near the characteristic line. *Comm. Pure Appl. Math.*, 63(8):1017–1070, 2010.

[9] Jinho Baik and Eric M. Rains. Limiting distributions for a polynuclear growth model with external sources. *J. Statist. Phys.*, 100(3-4):523–541, 2000.

[10] Jinho Baik and Eric M. Rains. Symmetrized random permutations. In *Random matrix models and their applications*, volume 40 of *Math. Sci. Res. Inst. Publ.*, pages 1–19. Cambridge Univ. Press, Cambridge, 2001.

[11] M. Balázs, J. Quastel, and T. Seppäläinen. Fluctuation exponent of the KPZ/ stochastic Burgers equation. *J. Amer. Math. Soc.*, 24(3):683–708, 2011.

[12] A. L. Barabási and H. E. Stanley. *Fractal Concepts in Surface Growth*. Cambridge University Press, 1995.

[13] Michael Bernkopf. A history of infinite matrices. *Arch. History Exact Sci.*, 4(4):308–358, 1968.

[14] Lorenzo Bertini and Nicoletta Cancrini. The stochastic heat equation: Feynman-Kac formula and intermittence. *J. Statist. Phys.*, 78(5-6):1377–1401, 1995.

[15] Lorenzo Bertini and Giambattista Giacomin. Stochastic Burgers and KPZ equations from particle systems. *Comm. Math. Phys.*, 183(3):571–607, 1997.

[16] Lorenzo Bertini and Giambattista Giacomin. On the long-time behavior of the stochastic heat equation. *Probab. Theory Related Fields*, 114(3):279–289, 1999.

[17] Jean Bertoin. Some properties of Burgers turbulence with white or stable noise initial data. In *Lévy processes*, pages 267–279. Birkhäuser Boston, Boston, MA, 2001.

[18] Giulio Biroli, Jean-Philippe Bouchaud, and Marc Potters. Extreme value problems in random matrix theory and other disordered systems. 2007. cond-mat/0702244.

[19] Erwin Bolthausen. A note on the diffusion of directed polymers in a random environment. *Comm. Math. Phys.*, 123(4):529–534, 1989.

[20] Folkmar Bornemann, Patrik L. Ferrari, and Michael Prähofer. The Airy_1 process is not the limit of the largest eigenvalue in GOE matrix diffusion. *J. Stat. Phys.*, 133(3):405–415, 2008.

[21] Alexei Borodin and Ivan Corwin. Macdonald processes. *arXiv:1111.4408*, 11 2011.

[22] Alexei Borodin, Ivan Corwin, and Daniel Remenik. Log-gamma polymer free energy fluctuations via a Fredholm determinant identity. *arXiv:1206.4573*, 06 2012.

[23] Alexei Borodin, Patrik L. Ferrari, Michael Prähofer, and Tomohiro Sasamoto. Fluctuation properties of the TASEP with periodic initial configuration. *J. Stat. Phys.*, 129(5-6):1055–1080, 2007.

[24] Alexei Borodin, Patrik L. Ferrari, and Tomohiro Sasamoto. Transition between Airy_1 and Airy_2 processes and TASEP fluctuations. *Comm. Pure Appl. Math.*, 61(11):1603–1629, 2008.

[25] Alexei Borodin, Patrik L. Ferrari, and Tomohiro Sasamoto. Two speed TASEP. *J. Stat. Phys.*, 137(5-6):936–977, 2009.

[26] Éric Brunet and Bernard Derrida. Probability distribution of the free energy of a directed polymer in a random medium. *Phys. Rev. E (3)*, 61(6, part B):6789–6801, 2000.

[27] Pasquale Calabrese and Pierre Le Doussal. An exact solution for the kpz equation with flat initial conditions. *Phys. Rev. Lett.*, 106:250603, 2011.

[28] Pasquale Calabrese, Pierre Le Doussal, and Alberto Rosso. Free-energy distribution of the directed polymer at high temperature. *2010 EPL 90 20002*, 02 2010.

[29] Philippe Carmona and Yueyun Hu. On the partition function of a directed polymer in a Gaussian random environment. *Probab. Theory Related Fields*, 124(3):431–457, 2002.

[30] Eric Cator and Leandro P. R. Pimentel. Hydrodynamical methods in last passage percolation models. *Publicações do 28o CBM - 2011*, 06 2011.

[31] T. Chan. Scaling limits of Wick ordered KPZ equation. *Comm. Math. Phys.*, 209(2):671–690, 2000.

[32] Sourav Chatterjee. The universal relation between scaling exponents in first-passage percolation. arXiv:1105.4566, 2011.

[33] Francis Comets, Tokuzo Shiga, and Nobuo Yoshida. Directed polymers in a random environment: path localization and strong disorder. *Bernoulli*, 9(4):705–723, 2003.

[34] Francis Comets, Tokuzo Shiga, and Nobuo Yoshida. Probabilistic analysis of directed polymers in a random environment: a review. In *Stochastic analysis on large scale interacting systems*, volume 39 of *Adv. Stud. Pure Math.*, pages 115–142. Math. Soc. Japan, Tokyo, 2004.

[35] Francis Comets and Nobuo Yoshida. Brownian directed polymers in random environment. *Comm. Math. Phys.*, 254(2):257–287, 2005.

[36] Francis Comets and Nobuo Yoshida. Directed polymers in random environment are diffusive at weak disorder. *Ann. Probab.*, 34(5):1746–1770, 2006.

[37] Ivan Corwin. The kardar-parisi-zhang equation and universality class. 06 2011.

[38] Ivan Corwin, P.L. Ferrari, and S. Peche. Universality of slow decorrelation in kpz growth. 01 2010.

[39] Ivan Corwin and Alan Hammond. Brownian gibbs property for airy line ensembles. arXiv:1108.2291, 2011.

[40] Ivan Corwin and Alan Hammond. Brownian gibbs property for airy line ensembles. 08 2011.

[41] Ivan Corwin, Neil O'Connell, Timo Seppäläinen, and Nikos Zygouras. Tropical combinatorics and Whittaker functions. 10 2011.

[42] Ivan Corwin and Jeremy Quastel. Crossover distributions at the edge of the rarefaction fan. arXiv:1006.1338, to appear in Ann. Prob., 2011.

[43] Ivan Corwin and Jeremy Quastel. Renormalization fixed point of the KPZ universality class. 03 2011.

[44] Giuseppe Da Prato and Jerzy Zabczyk. *Stochastic equations in infinite dimensions*, volume 44 of *Encyclopedia of Mathematics and its Applications*. Cambridge University Press, Cambridge, 1992.

[45] Robert Dalang, Davar Khoshnevisan, Carl Mueller, David Nualart, and Yimin Xiao. *A minicourse on stochastic partial differential equations*, volume 1962 of *Lecture Notes in Mathematics*. Springer-Verlag, Berlin, 2009. Held at the University of Utah,

Salt Lake City, UT, May 8–19, 2006, Edited by Khoshnevisan and Firas Rassoul-Agha.
[46] B. Derrida. Directed polymers in a random medium. *Physica A: Statistical and Theoretical Physics*, 163(1):71–84, 1990.
[47] Bernard Derrida and Antoine Gerschenfeld. Current fluctuations of the one dimensional symmetric simple exclusion process with step initial condition. *J. Stat. Phys.*, 136(1):1–15, 2009.
[48] Victor Dotsenko. Bethe anzats derivation of the tracy-widom distribution for one-dimensional directed polymers. *Eutrophys.Lett.*, v.90:20003, 2010.
[49] Victor Dotsenko. Replica bethe ansatz derivation of the tracy-widom distribution of the free energy fluctuations in one-dimensional directed polymers. *J. Stat. Mech. P07010 (2010)*, 04 2010.
[50] Victor Dotsenko and Boris Klumov. Attempted bethe ansatz solution for one-dimensional directed polymers in random media. *J. Stat.*, Mech.:P03022, 2010.
[51] Pierre Le Doussal and Pasquale Calabrese. The kpz equation with flat initial condition and the directed polymer with one free end. *J. Stat.*, Mech.:P06001, 2012.
[52] Freeman J. Dyson. A Brownian-motion model for the eigenvalues of a random matrix. *J. Mathematical Phys.*, 3:1191–1198, 1962.
[53] Patrik L. Ferrari. From interacting particle systems to random matrices. *J. Stat. Mech. Theory Exp.*, (10):P10016, 15, 2010.
[54] Patrik L. Ferrari and Herbert Spohn. A determinantal formula for the GOE Tracy-Widom distribution. *J. Phys. A*, 38(33):L557–L561, 2005.
[55] Patrik L. Ferrari and Herbert Spohn. Scaling limit for the space-time covariance of the stationary totally asymmetric simple exclusion process. *Comm. Math. Phys.*, 265(1):1–44, 2006.
[56] Patrik L. Ferrari and Herbert Spohn. Random growth models. arXiv:1003.0881, 2010.
[57] Daniel S. Fisher and David A. Huse. Directed paths in a random potential. *Phys. Rev. B*, 43(13):10728–10742, May 1991.
[58] Michael E. Fisher. Walks, walls, wetting, and melting. *J. Statist. Phys.*, 34(5-6):667–729, 1984.
[59] P. J. Forrester, T. Nagao, and G. Honner. Correlations for the orthogonal-unitary and symplectic-unitary transitions at the hard and soft edges. *Nuclear Physics B*, 553(3):601–643, 1999.
[60] Dieter Forster, David R. Nelson, and Michael J. Stephen. Large-distance and long-time properties of a randomly stirred fluid. *Phys. Rev. A (3)*, 16(2):732–749, 1977.
[61] Dieter Forster, David R. Nelson, and Michael J. Stephen. Large-distance and long-time properties of a randomly stirred fluid. *Phys. Rev. A*, 16(2):732–749, Aug 1977.
[62] L. Frachebourg and Ph. A. Martin. Exact statistical properties of the Burgers equation. *J. Fluid Mech.*, 417:323–349, 2000.
[63] Tadahisa Funaki and Jeremy Quastel. Invariant measures for the stochastic heat equation. In preparation.
[64] Adriano M. Garsia. Continuity properties of Gaussian processes with multidimensional time parameter. In *Proceedings of the Sixth Berkeley Symposium on Mathematical Statistics and Probability (Univ. California, Berkeley, Calif., 1970/1971), Vol. II: Probability theory*, pages 369–374, Berkeley, Calif., 1972. Univ. California Press.
[65] Jürgen Gärtner. Convergence towards Burgers' equation and propagation of chaos for weakly asymmetric exclusion processes. *Stochastic Process. Appl.*, 27(2):233–260, 1988.
[66] Patricia Goncalves and Milton Jara. Universality of KPZ equation. arXiv:1003.4478, 2010.
[67] Martin Hairer. Solving the KPZ equation. arXiv:1109.6811, 2011.

[68] O. Hallatschek, P. Hersen, S. Ramanathan, and D. R. Nelson. Genetic drift at expanding frontiers promotes gene segregation. *Proc. Natl. Acad. Sci.*, 104:19926–19930, 2007.

[69] Timothy Halpin-Healy and Yi-Cheng Zhang. Kinetic roughening phenomena, stochastic growth, directed polymers and all that. aspects of multidisciplinary statistical mechanics. *Physics Reports*, 254:215–414, 1995.

[70] Helge Holden, Bernt Øksendal, Jan Ubøe, and Tusheng Zhang. *Stochastic partial differential equations*. Universitext. Springer, New York, second edition, 2010. A modeling, white noise functional approach.

[71] David A. Huse and Christopher L. Henley. Pinning and roughening of domain walls in Ising systems due to random impurities. *Phys. Rev. Lett.*, 54(25):2708–2711, Jun 1985.

[72] T. Imamura and T. Sasamoto. Fluctuations of the one-dimensional polynuclear growth model with external sources. *Nuclear Phys. B*, 699(3):503–544, 2004.

[73] J. Z. Imbrie and T. Spencer. Diffusion of directed polymers in a random environment. *J. Statist. Phys.*, 52(3-4):609–626, 1988.

[74] I. Iscoe. On the supports of measure-valued critical branching Brownian motion. *Ann. Probab.*, 16(1):200–221, 1988.

[75] Svante Janson. *Gaussian Hilbert spaces*, volume 129 of *Cambridge Tracts in Mathematics*. Cambridge University Press, Cambridge, 1997.

[76] Kurt Johansson. Shape fluctuations and random matrices. *Comm. Math. Phys.*, 209(2):437–476, 2000.

[77] Kurt Johansson. Discrete polynuclear growth and determinantal processes. *Comm. Math. Phys.*, 242(1-2):277–329, 2003.

[78] Kurt Johansson. Random matrices and determinantal processes. In *Mathematical statistical physics*, pages 1–55. Elsevier B. V., Amsterdam, 2006.

[79] Mehran Kardar. Lectures on directed paths in random media. In *Géométries fluctuantes en mécanique statistique et en théorie des champs (Les Houches, 1994)*, pages 1–75. North-Holland, Amsterdam, 1996.

[80] Mehran Kardar, Giorgio Parisi, and Yi-Cheng Zhang. Dynamic scaling of growing interfaces. *Phys. Rev. Lett.*, 56(9):889–892, Mar 1986.

[81] Yuri Kifer. The Burgers equation with a random force and a general model for directed polymers in random environments. *Probab. Theory Related Fields*, 108(1):29–65, 1997.

[82] N. Konno and T. Shiga. Stochastic partial differential equations for some measure-valued diffusions. *Probab. Theory Related Fields*, 79(2):201–225, 1988.

[83] Thomas Kriecherbauer and Joachim Krug. A pedestrian's view on interacting particle systems, KPZ universality and random matrices. *J. Phys. A*, 43(40):403001, 41, 2010.

[84] J. Krug and H. Spohn. Kinetic roughening of growing interfaces. In *Solids far from Equilibrium: Growth, Morphology and Defects*, pages 479–582. Cambridge University Press, Boston, MA, 1992.

[85] Hubert Lacoin. New bounds for the free energy of directed polymers in dimension 1+1 and 1+2. arXiv:0901.0699, 2009.

[86] Thomas M. Liggett. *Interacting particle systems*, volume 276 of *Grundlehren der Mathematischen Wissenschaften [Fundamental Principles of Mathematical Sciences]*. Springer-Verlag, New York, 1985.

[87] A. M. S. Macêdo. Universal parametric correlations at the soft edge of the spectrum of random matrix ensembles. *EPL (Europhysics Letters)*, 26(9):641, 1994.

[88] Satya N. Majumdar and Sergei Nechaev. Anisotropic ballistic deposition model with links to the Ulam problem and the Tracy-Widom distribution. *Phys. Rev. E (3)*, 69(1):011103, 5, 2004.

[89] Olivier Mejane. Upper bound of a volume exponent for directed polymers in a random environment. *Ann. Inst. H. Poincaré Probab. Statist.*, 40(3):299–308, 2004.
[90] Govind Menon and Ravi Srinivasan. Kinetic theory and Lax equations for shock clustering and Burgers turbulence. *J. Stat. Phys.*, 140(6):1195–1223, 2010.
[91] C. Mueller and R. Nualart. Regularity of the density for the stochastic heat equation.
[92] Carl Mueller. On the support of solutions to the heat equation with noise. *Stochastics Stochastics Rep.*, 37(4):225–245, 1991.
[93] Carl Mueller, Leonid Mytnik, and Jeremy Quastel. Effect of noise on front propagation in reaction-diffusion equations of KPP type. *Invent. Math.*, 184(2):405–453, 2011.
[94] N. J. Zabusky and M. D. Kruskal. Exact statistical properties of the Burgers equation. *Phys. Rev. Lett.*, 15:240–243, 1965.
[95] Neil O'Connell. Directed polymers and the quantum toda lattice. *Annals of Probability 2012*, Vol. 40, No. 2, 437–458, 10, 2009.
[96] Neil O'Connell and Jon Warren. A multi-layer extension of the stochastic heat equation. arXiv:1104.3509, 2011.
[97] É. Pardoux. Stochastic partial differential equations, a review. *Bull. Sci. Math.*, 117(1):29–47, 1993.
[98] E. Pardoux and T. Zhang. Absolute continuity of the law of the solution of a parabolic spde. *J. Functional Anal.*, 112, 1993.
[99] Markus Petermann. Superdiffusivity of directed polymers in random environment. PhD Thesis, University of Zürich, 2000.
[100] Michael Prähofer and Herbert Spohn. Statistical self-similarity of one-dimensional growth processes. *Phys. A*, 279(1-4):342–352, 2000. Statistical mechanics: from rigorous results to applications.
[101] Michael Prähofer and Herbert Spohn. Scale invariance of the PNG droplet and the Airy process. *J. Statist. Phys.*, 108(5-6):1071–1106, 2002. Dedicated to David Ruelle and Yasha Sinai on the occasion of their 65th birthdays.
[102] Michael Prähofer and Herbert Spohn. Scale invariance of the PNG droplet and the Airy process. *J. Statist. Phys.*, 108(5-6):1071–1106, 2002.
[103] Jeremy Quastel and Daniel Remenik. Local brownian property of the narrow wedge solution of the KPZ equation. arXiv:1105.0952 To appear in Elec. Comm. Prob., 2011.
[104] Jeremy Quastel and Benedek Valko. $t^{1/3}$ Superdiffusivity of finite-range asymmetric exclusion processes on $\mathbb{Z}$. *Comm. Math. Phys.*, 273(2):379–394, 2007.
[105] T. Sasamoto. Spatial correlations of the 1D KPZ surface on a flat substrate. *J. Phys. A*, 38(33):L549–L556, 2005.
[106] T. Sasamoto. Exact results for the 1D asymmetric exclusion process and KPZ fluctuations. *Eur. Phys. J. B*, 64(3-4):373–377, 2008.
[107] T. Sasamoto and H. Spohn. One-dimensional KPZ equation: an exact solution and its universality. *Phys. Rev. Lett.*, 104(23), 2010.
[108] Tomohiro Sasamoto and Herbert Spohn. Exact height distributions for the KPZ equation with narrow wedge initial condition. *Nuclear Phys. B*, 834(3):523–542, 2010.
[109] Timo Seppäläinen. Scaling for a one-dimensional directed polymer with boundary conditions. *Ann. Probab.*, 40(1):19–73, 2012.
[110] Barry Simon. *Trace ideals and their applications*, volume 120 of *Mathematical Surveys and Monographs*. American Mathematical Society, Providence, RI, second edition, 2005.
[111] Herbert Spohn. Kpz scaling theory and the semi-discrete directed polymer model. 01 2012.
[112] Kazumasa A. Takeuchi and Masaki Sano. Universal fluctuations of growing interfaces: Evidence in turbulent liquid crystals. *Phys. Rev. Lett.*, 104:230601, 2010.

[113] Kazumasa A. Takeuchi, Masaki Sano, Tomohiro Sasamoto, and Herbert Spohn. Growing interfaces uncover universal fluctuations behind scale invariance. *Sci. Rep. (Nature)*, 1:34, 2011.

[114] Craig A. Tracy and Harold Widom. A Fredholm determinant representation in ASEP. *J. Stat. Phys.*, 132(2):291–300, 2008.

[115] Craig A. Tracy and Harold Widom. Integral formulas for the asymmetric simple exclusion process. *Comm. Math. Phys.*, 279(3):815–844, 2008.

[116] Craig A. Tracy and Harold Widom. Asymptotics in ASEP with step initial condition. *Comm. Math. Phys.*, 290(1):129–154, 2009.

[117] Craig A. Tracy and Harold Widom. On ASEP with step Bernoulli initial condition. *J. Stat. Phys.*, 137(5-6):825–838, 2009.

[118] H. van Beijeren. Excess noise for driven diffusive systems. *Phys. Rev. Lett.; Physical Review Letters*, 54(18):2026–2029, 1985.

[119] John B. Walsh. An introduction to stochastic partial differential equations. In *École d'été de probabilités de Saint-Flour, XIV—1984*, volume 1180 of *Lecture Notes in Math.*, pages 265–439. Springer, Berlin, 1986.

[120] Wojbor A. Woyczyński. *Burgers-KPZ turbulence*, volume 1700 of *Lecture Notes in Mathematics*. Springer-Verlag, Berlin, 1998. Göttingen lectures.

[121] Mario V. Wüthrich. Fluctuation results for Brownian motion in a Poissonian potential. *Ann. Inst. H. Poincaré Probab. Statist.*, 34(3):279–308, 1998.

[122] Mario V. Wüthrich. Superdiffusive behavior of two-dimensional Brownian motion in a Poissonian potential. *Ann. Probab.*, 26(3):1000–1015, 1998.

Jeremy Quastel, Departments of Mathematics and Statistics, University of Toronto, 40 St George Street, Toronto, ON M5S 1L2

E-mail address: quastel@math.toronto.edu

Made in United States
North Haven, CT
05 November 2026

21764589R00109